Application Development

Application Development

Accelerated Lotus Notes Study Guide

Libby Ingrassia Schwarz

McGraw-Hill
New York • San Francisco • Washington, D.C. • Auckland
Bogotá • Caracas • Lisbon • London • Madrid • Mexico City
Milan • Montreal • New Delhi • San Juan • Singapore
Sydney • Tokyo • Toronto

Library of Congress Cataloging-in-Publication Data

Schwarz, Libby Ingrassia.
Accelerated Lotus Notes application development study guide / Libby Ingrassia Schwarz.
p. cm.
Includes index.
ISBN 0-07-134569-8
1. Lotus Notes for Windows—Examinations. 2. Business—Computer programs—Examinations. 3. Database management—Computer programs—Examinations. 4. Electronic mail systems—Examinations. 5. Groupware (Computer software)—Examinations. I. Title.
HF5548.4.L692S39 1998
005.369—dc21 98-43762
CIP

McGraw-Hill

A Division of The McGraw-Hill Companies

1 2 3 4 5 6 7 8 9 0 AGM/AGM 9 0 3 2 1 0 9 8

ISBN 0-07-134569-8

The sponsoring editor for this book was Judy Brief and the production supervisor was Clare Stanley. It was set by D & G Limited, LLC.

Printed and bound by Quebecor/Martinsburg.

Dedication

To my dad, Sal, for making me want to make you proud.

To my mom, Melinda, for being the best mom and friend a girl could ask for.

And, of course, to my wonderful Prince Charming, my husband Brian. You are the inspiration, the support, the joy, and the strength . . . Everything I am is better with you!

Acknowledgments

No book is ever written by one person alone, with no help from others. This book is, of course, no exception.

I would like to thank my technical editors, artists, and supporters: Michael Meyers and Scott Jernigan. Without them, this book would never have made it to completion!

All of my other friends at Total Seminars also deserve a huge round of thanks for making it possible for me to write—and for putting up with me while I did.

Also, many thanks to Judy Brief, my editor at McGraw-Hill, for giving me the opportunity to write this book.

Finally, love and appreciation to all my friends, my family, and my husband, who are all wondering, "Who is that woman?" now that I have emerged from the depths of writing.

Contents

INTRODUCTION

Introduction to Lotus Notes

When you purchase Lotus Notes books or tell people that you are studying for a Lotus Notes certification examination, the first question you are asked is, "What exactly *is* Lotus Notes, anyway? Some sort of mail client, right?" This is when you begin your sideline career of explaining Lotus Notes and the concept of groupware.

Lotus Notes markets itself as the premier groupware product on the market, competing with other programs such as Microsoft Exchange and Novell GroupWise. *Groupware* is any software that enhances the abilities of groups of people (including departments and workgroups) to communicate and conduct business with each other. Groupware usually offers multiple functions, including e-mail, knowledge management, and collaboration. Though Lotus Notes functions quite well as strictly an e-mail engine, Notes can do much more. Notes can also perform a variety of other workgroup communication functions, including sharing documents, automating business processes using workflow applications, publishing on the World Wide Web, creating internal corporate intranets and more.

Here are two examples of how we define Notes by describing the program's functions in an office environment. First, Notes works much like the discussion groups on the Internet. In a discussion group, users can write a message containing information and share it with multiple people, without having to e-mail the message to everyone. This is useful for two reasons. First, sharing information without using e-mail saves an amazing amount of space on servers. Second, when people leave or join an organization or workgroup, information can be lost or inefficiently passed on to others. It is easier to give a new user in a workgroup the rights to a database that contains all the shared data on a project, rather than to forward all the e-mail related to the project.

Lotus Notes significantly reduces both the amount of paper used in an office and the time necessary to finalize a request through multiple departments. One of the examples I give as an instructor, in trying to describe this feature of Notes, is to ask one of my students to describe a business process in their organization, such as asking for time off or requesting expense reimbursements. Students usually describe a process that requires filling out forms, then walking to a department or sending the form to the appropriate person for approval. That person would then forward the form to three or four more people before the process was completed. Approving the form often included multiple people, with several places in which the form could be sidetracked or lost, often without any way of recovery. Notes automates such processes and allows a person to track the progress of a request, to find out who has signed it and who has not, and even to send a memo to the latter to nudge them into action. A Notes developer can customize the program to accomplish just about any task in an office setting. In the example, a Lotus Notes developer could create the request form in a database. Once the database is completed and saved as a document, the developer could place that form in a database view where the approval process will take place. This leaves an electronic paper trail for those involved in the process, so they can track the status and location of the form. As a Notes Application Developer, you will create databases to fulfill the business needs of your organization. You may also have to customize, troubleshoot or update existing applications and databases. As a candidate for certification, you are expected to develop Notes applications and troubleshoot common problems that occur with Notes applications.

Introduction to This Book

Certifications from software and hardware vendors are more than popular in the information technology industry—they have become necessary. These certifications allow companies to hire employees or consultants who have proven their knowledge of the products they will implement or support.

Like many other vendors, Lotus implements a set of vendor certifications for Lotus Notes. To obtain these certifications, you need knowledge of the Domino and Notes environments—and you need to prove your skills by taking the Lotus Education exams. These certifications are valuable for many reasons. First, salary surveys in a variety of magazines consistently show higher salaries for certified employees. Also, for those who are trying to find their niche in the busy information technology industry, a certification is often necessary for obtaining that all-important first interview or job.

Everyone has different methods of preparing for certification exams. Some take the official Lotus Education courses offered by Lotus Authorized Education Centers. Others attempt to study on their own or rely on their own knowledge of the software. Others search local bookstores for a guide to passing the exams. While there are many good books on the market to assist in designing and implementing a Notes application or environment, there are relatively few Notes certification study guides.

This book is a Notes certification study guide. It is intended to be a quick reference to the information covered on the Application Development I and II exams. No matter what other types of studying you choose to do, this book will help you refine your study to concentrate on the material actually covered on the exams.

The goal of this book, therefore, is to help you pass the exams necessary to become certified in Lotus Notes Application Development. Specifically, the book is intended to prepare you for passing the multiple choice Application Development I and Application Development II exams, offered by Lotus through its independent testing vendors.

To that end, each section of the book displays objectives that relate to questions at the end of each section. These objectives are based on the lists of competencies published by Lotus Education in its certification exam guides. The objectives also are based on personal experience with taking the Lotus exams, teaching the official Lotus courses, and implementing Notes enterprise environments for a variety of companies.

To help you study, I recommend reading each chapter first, looking for information required to understand the objectives as you read the text. Then, you should answer the questions at the end of each section immediately after reading that particular section. Later in your studying, go through the chapters again. This time, read the bulleted lists and definitions. You also want to look over the screen captures and graphics again. Then go over the questions again to review the concepts and make sure you have not forgotten anything. You will notice that certain concepts and details are covered more than once in the text. This is not an insult to your intelligence. It is one way to ensure that you have the opportunity to review important topics multiple times in the course of your studying. The topics that you see frequently in the text are covered in depth in the courses and official certification guides. You also should expect to see them covered thoroughly on the exams.

Besides reading this book and answering the practice questions in each section, I would highly recommend that you work with Lotus Notes. For most of the exercises in Application Development I, you only need access to a Notes client. To start developing, you will need an ID that has a regular Notes license. When you decide to set up a server to create your ID and to work on some of the more advanced topics in Application Development II that require a server, keep in mind that you do not need any fancy hardware to set up a Notes server. It can run on almost any operating system. If you do not already work with Notes, you can obtain a trial version from Notes Net at `http://notes.net`. The exam designers assume that you have experience developing a database, creating forms and views, and writing and debugging formulas. You simply cannot get this experience from any other source. This book assumes, moreover, that you have some experience with Lotus Notes—enough, at least, to have some familiarity with the functions and feel of the software.

Lotus Certification Paths and Resources

Certifications awarded by Lotus for Notes Releases 4.X include: *Certified Lotus Specialist* (CLS), *Certified Lotus Professional* (CLP), *Principal Certified Lotus Professional* (Principal CLP), Domino Messaging Administrator (R4.6 only), and *Certified Lotus Instructor* (CLI). Exams covered in this book are required for CLP, Principal CLP, and CLI for Notes Releases 4.0 to 4.5. For Notes Releases 4.6 and higher, Lotus has released additional exams and certifications (see the exam matrix for more information). The Application Development I exam is an exam option for CLS status. When you combine Application Development I and II with a System Administration I exam (covered in another McGraw-Hill Accelerated Guide), you will be a Certified Lotus Professional—Application Developer.

Becoming a CLS requires you to pass one examination. Exams required for CLS are displayed in the Exam Matrix table. Becoming a CLP, however, requires three exams. The two basic, required exams are System Administration I and Application Development I. For the third exam, the certification candidate chooses whether to specialize in System Administration or Application Development. According to Lotus statistics, there are approximately 60 percent application developers and 40 percent system administrators in the CLP mix.

NOTE

The following exam matrix lists only some of the possible examination paths. Other exams exist that qualify a candidate for CLS certification, such as the Domino Web Development and Administration exam, the Developing LotusScript Applications for SmartSuite exam and the cc:Mail exam. Also, there are new tracks available for Notes release 4.6 certifications, including the Domino Messaging Administrator and the alternative exams available for CLP certifications. For updated information on these alternatives, see the Lotus Education and Certification Web site at the following URL:

```
http://www.lotus.com/education
```

Exam Matrix

Exams for CLS status (4.5):	Add these exams for CLP Application Developer (4.5):	Add these exams for Principal CLP Application Developer (4.5):	Add these exams for CLP System Administrator (4.5):	Add these exams for Principal CLP System Administrator (4.5):
Application Development I (190–271)	System Administration I (190–274); Application Development II (190–272)	LotusScript in Notes for Advanced Developers (LND) (190–273); Developing Domino Applications for the Web (DDA) (190–278); Domino Web Development and Administration (DWDA) (190–281); Developing LotusScript for SmartSuite 1-2-3 Apps (DLAS) (190–291)	System Administration I (190–274); System Administration II (190-275)	Administrating Specialized Tasks for Domino 4.5 (AST) (190–276); cc:Mail R6 System Administration 1 (190–251)

Exams for CLS status (4.5):	Add these exams for CLP Application Developer (4.5):	Add these exams for Principal CLP Application Developer (4.5):	Add these exams for CLP System Administrator (4.5):	Add these exams for Principal CLP System Administrator (4.5):
System Administration I (190–274)	Application Development I (190–271); Application Development II (190–272)		Application Development I (190–271); System Administration II (190–275)	

Getting More Information

Like all Web-based information printed in books, the Lotus exams and certifications described in the matrix may have changed somewhat by the time you read this. To ensure that you have updated information, refer to the following sources as you start down your certification path—and as you continue to progress.

- Lotus Education Web Page—Lotus' Web site, at `http://www.lotus.com`, is a rich source of additional information about exams, exam topics and competencies, and certifications. Specifically, if you go to the Education home page at `http://www.lotus.com/education`, you will find updated information about the exams and the exam matrix, as well as updates on the certification paths themselves. You also can find the official Lotus exam guides, which you can download in Adobe Acrobat file format (.pdf).
- Lotus Education Helpline—If you have specific questions about taking the exams or taking official Lotus courses, or if you want to know more about Lotus certifications, call the Lotus Education Helpline at (800) 346-6409 or (617) 693-4436. The Helpline is open from 8:30 a.m. to 5:30 p.m. ET, Monday through Friday.

The Exams

After you determine the path for your Lotus Notes certifications, choose your first exam and begin studying. If you purchased this book, I assume that you made some of these decisions, such as which exams to take. You must also choose which testing format you prefer: multiple choice or concurrent application.

Lotus offers two types of examination formats available for the Level One exams. For all Lotus exams, there is the typical multiple choice examination. For both Application Development I and System Administration I, however, Lotus also offers concurrent application testing. This type of exam offers the person taking the exam an opportunity to work with Lotus Notes and Domino to accomplish specific tasks during the allotted time. Lotus intends to offer the concurrent application format for the Application Development II and System Administration II exams by the end of 1998.

The multiple choice tests require the person taking the exam to answer approximately 40 to 60 questions in multiple choice format, including true or false questions, questions with multiple answers, and questions that require users to interpret graphics or diagrams. Each test is computer-based and timed, with the time limit set specifically for each exam. The testing proctor will ensure that you know the time limit, and the test interface will show you how much time you have left during the test. Each test is closed-book. You will be given blank scratch paper (which must be returned to the testing proctor after the exam) and a pen or pencil if you need it. No other items may be brought into the testing center, including purses, pagers, notes, books, and calculators. At the beginning of the exam, the testing interface will display the exact number of questions on the exam and the passing percentage. When you complete the test, the interface will display your percentage correct and the passing percentage. This also will be printed at the testing center for you to keep and will be forwarded to Lotus Education within five business days. If you do not pass one of the exams, you must register and pay the exam fee again through Sylvan Prometric or CATGlobal.

NOTE
Testing rules for each center are explained to the candidate both at registration and at testing time. Be aware that Lotus and the testing centers take these rules seriously. If a candidate is accused of and proven to have passed an exam through any inappropriate or questionable means, the candidate's scores will be dismissed and the candidate will have to retest after a six-month waiting period. The candidate can take no Lotus certification exams within the six-month waiting period.

This book specifically targets the multiple choice type of examination. The information and competencies covered, however, apply to both the multiple choice and the concurrent application testing formats. Working through the sample database offered in each chapter, as well as the review questions, should prepare you for the concurrent application testing format—if that is the format you prefer. More hands-on work with Notes is highly recommended for testing using the concurrent application method.

The next step in pursuing the Notes certification is to study for and take the exams you have selected. Obviously, everyone learns and studies differently. This book is an excellent study tool for all types of learning methods, in that I have tried to give you the chance to read about the topics you need to know and to work through some of the tasks that are required competencies. Besides reading this book and working through the sample databases and review questions, you may find that Lotus Learning Bytes, an unofficial course available at `http://notes.net/lbytes.nsf/`, computer-based training, and a variety of practice exams will help you pass the exams.

Many find that hands-on experience is the best way to learn software and pass exams. One way to receive this type of experience is to take classes in Notes. Two sets of classes are available to help you pass the Notes certification exams. First, there are the official Lotus Curriculum technical and end-user classes, which are hands-on, instructor-led courses. The technical courses usually last two to five days while the end-user courses last a half-day. You will find many of the same topics from this book covered in the courses. Note, however, that Lotus' stated objective for students in the classes is learning the tasks related to their jobs—such as learning to be good application developers. The courses are refined somewhat differently than the exams, in my experience. The official courseware, in addition, has been updated more recently than the exams. The other classes available include several unofficial instructor-led courses and a variety of computer-based training (CBT) alternatives. These include Lotus Learning Bytes and CD-ROM-based CBT—again, from multiple companies. After using all these methods to study, you may also want to buy some of the practice exams offered on the market. The practice exams not only give you the ability to test your knowledge, but they also allow you to practice taking an exam, including time limits and computerized testing. Mastering these skills can be a huge help in passing the exams. Finally, as mentioned before, you can obtain a trial version of Lotus Notes and Domino from `http://notes.net`. This gives you the chance to practice all the skills and techniques necessary to pass the exams.

HOW LONG SHOULD I STUDY?

After helping many individuals through the Application Development I and II exams, I developed an idea of how long it takes to master the material and pass the exams. The time it takes to study for the exams is described in the following table. The number of hours listed in the table only represent averages. If you are a slower, more methodical person, or if you are a nervous test-taker, add 10 percent. If, however, you are the person who mastered chemistry in one night and are extremely methodical, you may want to reduce the numbers by 10 percent. To use the table, circle the values under the column that represent how much time you have spent doing the tasks listed. Then add the values to get a total number of study hours.

- If you are new to Notes in general, add 20 hours to the total.
- If you are comfortable with Notes as an end user or system administrator, add 10 hours to the total.
- If you have been a Notes administrator, subtract five hours from the total.
- If you have been a Notes developer, subtract 15 hours from the total.
- If you are a developer in another type of environment but are not familiar with Notes, you should add 15 hours to the total.

Amount of experience doing . . .	None	Once or Twice	Occasionally	Often
Used the Notes client	10	8	7	5
Created a Notes database	10	8	7	5
Created Notes forms	8	6	4	3
Used subforms	6	4	2	1
Used Hide-When	10	8	6	4
Created fields on Notes forms	6	4	2	1
Used Notes @functions	8	6	4	2
Used Notes @commands	4	3	2	1
Created Actions, Buttons and Hotspots	6	4	2	1
Created Agents	4	3	2	1
Created Notes views	10	8	6	4
Used selection formulas	5	4	2	1
Created Notes database ACLs	6	4	2	1
Edited About and Using documents	4	3	2	1

continues

Amount of experience doing . . .	None	Once or Twice	Occasionally	Often
Created a Navigator	2	1	1	1
Used workflow and mailing formulas	8	6	4	2
Used layout regions	6	4	2	1
Used computed subforms	4	3	2	1
Used environment variables in formulas	4	3	2	1
Used Notes internal and reserved fields	6	4	2	1
Used @DBLookUp and @DBColumn	8	6	4	2
Used @PickList and @DialogBox	6	4	2	1
Used form formulas in views	6	4	2	1
Created scheduled and automatic agents	4	3	2	1
Used templates to maintain databases	6	4	2	1
Used reader and author fields	6	4	2	1
Used form and view access lists	6	4	2	1
Used signing and encryption	6	4	2	1
Created and used User Roles	4	3	2	1

How to Take the Tests

The following paragraphs outline some test-taking strategies that often make the multiple choice-style exams easier.

1. Use the time before you begin the test to jot down any memorized notes on the scratch paper provided.
2. Go through the exam once, answering any questions of which you are certain. Make notes of any questions giving you information that might help you answer another question.
3. Mark any questions you are not sure of and return to them on your second pass.
4. Answer *all* the questions. Blank questions are automatically considered wrong.
5. Sometimes it helps to write or draw part of the information given in a question. This especially helps clarify scenario questions.
6. Read each question slowly and carefully. Make sure you know what the question is asking before you answer. Do not go too quickly and miss a key word.
7. Visualize the Lotus Notes interface as clearly as you can, including menus and dialog boxes. Draw diagrams to help you, if necessary. Knowing where options are located and how to perform actions will be important.

Some of these strategies may work for you, while others may not. I recommend using practice exams to try out these ideas, to see which ones work for you. The more at ease you are with the testing format, the more you can concentrate on the questions. Good luck.

REGISTERING FOR THE EXAMS

After completing your studying, you can register for the exams by contacting one of Lotus' two independent testing vendors. For the multiple choice exams, register by contacting Sylvan Prometric testing centers at one of the following numbers:

```
(800) 74-LOTUS (which is 800-745-6887), or (612) 896-7000.
```

You may also register for exams online at Sylvan by going to Sylvan Prometric's Online Registration site at the following URL:

```
http://www.slspro.com/clients/Register/sls/aptc/testthrough.htm
```

You will need the following information ready, either over the phone or online, when you register:

- Name
- Social Security Number or Testing ID
- Mailing Address and Phone Number
- Company Name
- Name and Number of the Exam You Are Registering to Take
- When and Where You Wish to Test (you can find testing center information at the Sylvan Prometric Web site):

  ```
  http://www.sylvanprometric.com/nav.htm.
  ```

- Payment Method (credit card, voucher, money order or check). Note: You cannot test until they receive payment, which can postpone testing if paying by money order or by check.

To register for the concurrent application testing, contact CATGlobal Testing Centers. Complete your CATGlobal registration online at the following URL:

```
http://www.catglobal.com
```

CHAPTER 1

Notes Applications

Chapter Objectives

After reading this chapter, you should be able to answer questions based on the following objectives:

- Describe the types of Notes databases.
- Define appropriate uses for Notes applications.
- Compare Notes database applications to relational database applications.
- Understand options for the Database Properties InfoBox, including full-text indexing and launch options.
- Create a new database.
- Choose a size limit for a new database.

A Lotus Notes application is the combination of one or more Notes databases designed to store and share information that involves a particular business function. A Notes database is a container for multiple documents about related topics. These documents can contain rich text, graphics, tables, buttons, fields, formulas, and many other types of information. With the *View* feature of a database, users can scan each document quickly to determine whether that document has the needed data. A Notes application manipulates the various subsets of stored database information as a single entity. A request for a new box of paper, for example, would immediately generate an e-mail to the purchasing department. As soon as the order is filled, the inventory database will be updated. All this can be accomplished with one Notes application, using multiple Notes databases. Lotus Notes Application Development I explains how to create Notes databases. Application Development II, however, discusses how to combine multiple databases into a single Notes application.

Notes uses five different types of databases: broadcast, workflow, discussion, reference, and tracking. Each type of database stores and shares information differently and is, therefore, used for different purposes. The following paragraphs describe each of the five database types.

Broadcast databases are used for giving groups of users new information. Broadcast databases enable you to update users without necessarily sending e-mail messages. Newsletters and daily news releases are good uses of this type of application.

Workflow databases use a business process, or cycle, to move data. This type of application usually is used to route information through a request-and-approval cycle, which requires the input of many different users. This type of database can be used for vacation or purchase requests, for example.

Discussion databases are the simplest and most common types of Notes databases. These databases track main documents and responses to those documents, much like Internet newsgroups. It is useful for group brainstorming, suggestions, or any type of workgroup communication.

A *Reference* database, sometimes also called a Document Library database, stores data. An example of a reference database that you may already have used is the Help database installed with Notes. This type of database can be used for any type of documentation.

Finally, a *Tracking* database is used to monitor a process or performance. A tracking database, for example, might be used to identify people who check out a piece of software or hardware. It might also be useful for tracking reservations or sales.

An application often will use more than one type of database to fulfill the needs of a group of users. An application created for help desk call tracking, for example, would have a tracking database to monitor call tickets. It might also have an element of a workflow database to route the calls to the appropriate help desk personnel. The help desk application also would probably have an element of the reference database, to ensure that answers to questions previously asked would be available when those questions were asked again. This type of application would be a good use of Lotus Notes.

Some applications are not as appropriate for a Notes database and might be better served by a *relational* database, such as Microsoft Access. Some applications that would be better served by another type of software include the following:

- Applications that require real-time data transactions. Notes data is distributed through a periodic, scheduled replication process that can take time to propagate the information to all the servers in an enterprise.
- Applications that require data amounts larger than 4G. Lotus Notes databases have a hard-coded, 4G limit. Applications that require larger amounts of data would therefore be better suited to another application. Practical limits for applications may be even less than that, after factoring in the speed of the servers, workstations, and network. By default, the Notes database limit automatically set is 1G.
- Applications that require complex mathematical computations.
- Applications that require complex reports or ad-hoc queries. Notes does not support SQL queries or contain a report writer.

Lotus Notes databases differ internally from relational databases such as Microsoft Access. The two should not be confused. A Lotus Notes database creates and uses documents to store data; whereas in a relational database, the data is transactional and is stored in records. Table 1.1 describes a comparison between Notes databases and relational databases. Notes databases and applications do not replace relational database applications; rather, they can be complementary when the two types of applications support one another. A sales manager, for example, may want to fill out a monthly sales report form in a Notes database, which would pull numerical data from external spreadsheets (relational databases) that contain expense and sales figures.

Table 1.1 Relational Databases versus Notes Databases

Relational Databases	Notes Databases
Transactional processing	Document organization
Structured data	Semi-structured data (graphics, rich text)
Real-time	Periodic replication
Ad-hoc queries	Predefined queries (views)
Single enterprise	Multiple enterprises
SQL access	Full text searchable

Elements of a Notes Database

Four main elements exist in a Notes database:

1. Forms
2. Fields
3. Documents
4. Views

A *form* is the basic element of a Notes database. The form gives structure and function to the data and contains fields that will be completed to store the data. A form also contains elements such as static text, graphics, hotspots, and buttons.

A *field* is the element of the form that contains data. A field can be completed manually by user input or automatically by a Notes-calculated formula. Some of the many field types include text, rich text, keywords, names, and time and date information, to name a few.

A form with data in its fields is saved as a *document*, which is the most basic unit of information in a Notes database. All documents must be connected to a form. Most documents are displayed based on the form that was used to create them, so they will also include the static text of the form that actually defines or describes the data. Sometimes, the text that explains the field changes, based on whether you are creating or viewing the data. In other cases, data will be displayed on an entirely different form than the one used to create the data.

Finally, the *view* organizes and displays the data contained in the documents. Without views in a database, you have no way to access the documents that you created and saved. Views are predefined queries. They select particular documents to display, based on a selection formula that searches on a particular criteria, such as *form types* or *authors*, for example. The data in a view is organized into rows and columns that can be sorted, categorized, and totaled.

In addition, other elements such as actions, agents, navigators, security, and help documents complete a database. These elements may or may not be directly related to the information in the database. They are necessary, however, to achieve the appropriate function. All the elements of a database, including these additional features, will be discussed more fully in later chapters.

Creating a Notes Database

Now that we know about Notes applications and databases, the next step is to create a Notes database. Choose File . . . Database . . . New and the New Database dialog box will be displayed, as in Figure 1.1.

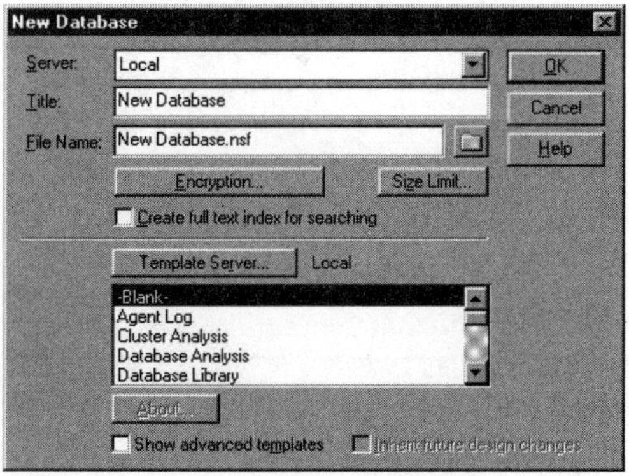

Figure 1.1 New Database dialog box.

The dialog box contains a series of fields that allow the database designer to customize the new database. Click OK after completing the fields. This creates and opens the database and places an icon on your workspace. Table 1.2 describes each of the fields in the New Database dialog box.

Table 1.2 New Database dialog box fields

Field	Value and Information
Server[1]	Default entry in this field is `Local`, which means your local machine (the location of your `\notes\data directory`). To place the new database on the server, either click the drop-down arrow to choose the appropriate server or type in the fully distinguished server name.
Title	This field is for the database title. Users will see this name on the database icon on their workspace. It should be descriptive of the database purpose. The maximum number of characters, including spaces, is 32.
Filename[2]	The filename field will automatically begin to fill in as the title is typed. It can be changed to suit the designer's preferences. Note that although long filenames are possible, they should be used with care if the database will reside on a server whose operating system does not support this feature. By default, the database will be stored in the `\notes\data` directory on the server you choose. To place the database in a subdirectory, click the Browse (folder) button to the right of the field. This enables you to choose a folder on the server or local machine.
Encryption button	This button allows local encryption in a database.
Size Limit	This button allows you to change the default size limit on databases. By default, the size limit is 1GB. You can change it to 2, 3 or 4GB. Once the database has been created, you cannot change the size unless you make a new replica.

Field	Value and Information
Template Server	**Click this button to choose the server that contains the database templates.**
Template List[3]	**Choose a template from the list to use as a base for your database, or choose** `Blank` **to start from scratch in building forms and views.**

[1]In most instances, it is best to create the Notes database on the workstation first. Then, when you are ready to test the database and implement some of the server-based functions, move the database to the server. If you created the database on the server, users might be able to open and use the database before it was completed.

[2]File extensions that are used in Notes include .nsf, .ntf, and .ns3. *Notes Storage Facility* (NSFs) are the default database extensions, which are added when you create a database. *Notes Template Facility* (NTFs) are Notes template extensions. NS3s are databases specifically compatible with Notes V3.

[3]New databases can be created in one of three ways: by copying an existing database, by basing a database on one of the available templates, or by creating an empty new database. The Application Development I exam centers on the person's ability to create a new, empty database.

Database templates contain design elements such as forms, views, actions, and navigators but do not contain any documents. Templates can be the basis for a new database, making it easier and faster to develop a new database. Use caution when basing a database on a template, however, because it can actually take longer to design if the views and forms are not truly appropriate for your purpose. The following is a list of some of the templates available with Notes Release 4.X:

- `APPROVE4.NTF`, used for an approval cycle database. This template creates a database that allows developers to add a standard approval cycle.
- `DISCUSS4.NTF`, used for a discussion database. This template creates a discussion area that can be used by members of workgroups to share their thoughts and ideas. This template includes archiving capabilities.
- `DOCLIB4.NTF`, used for a document library database. This template creates a library or file cabinet to store reference documents and includes design elements for a review cycle and for archiving

capabilities. Specific document libraries also are available for various software packages, including Lotus SmartSuite and Microsoft Office.

- `JOURNAL4.NTF`, used for a personal journal database. This template creates a database that serves as a diary for keeping ideas, documents and conversation or interview records. This template includes design elements for graphic navigators.
- `PROJECT4.NTF`, used for a project management database. This template creates a database that allows workgroups to keep track of project activities.
- `RESERVE45.NTF`, used for a room reservations database. This template creates a database that schedules and tracks the use of conference rooms, equipment or offices.

If you want to create one of these types of databases, or a database that uses the same design elements, using a template will make development easier and faster. Other types of templates available include the personal and public name and address book templates and the mail templates. These also can be used as a basis for new databases or for copying design elements. The Notes Help database provides more information on each of these template types.

Database Properties

After creating a database, go to the Database Properties InfoBox. Each element of a Notes database has a properties dialog box, also called an InfoBox. The Properties InfoBoxes enable you to create settings for most elements of your database. To access the Database Properties InfoBox, right-click on the database icon and choose Database Properties from the shortcut menu. Alternatively, you can choose File . . . Database . . . Properties. Notes separates Properties InfoBoxes into tabbed pages. The tabbed pages list settings for different elements of the database. Most properties InfoBoxes have six tabbed pages: Basic, Information, Print, Design, Launch and Full-Text tabs.

The Basics Tab

The Database Properties InfoBox Basic tab, which enables you to display and change the title, type of database, and other settings is shown in Figure 1.2. The Basic tab fields are described in Table 1.3. Other options may be visible on this tab, depending on your version of Notes. The ones listed may be tested on the Application Development I exam.

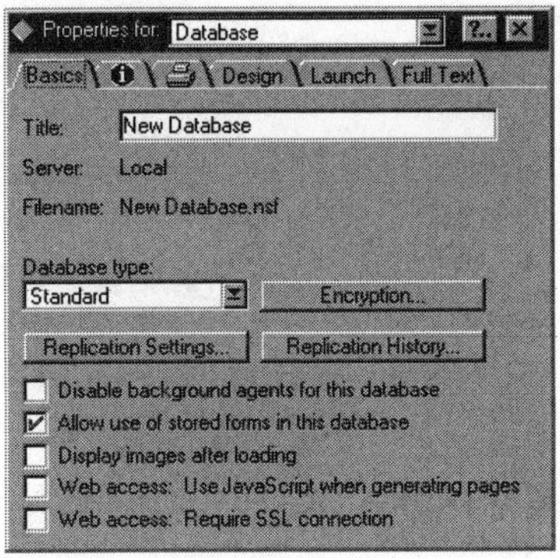

Figure 1.2 Database Properties InfoBox Basics tab.

Table 1.3 Database Properties InfoBox Basic Tab with Values and Information Given for the Fields

Field	Value and Information
Title	This is the title given to the database when it is created. You can change the title if you have manager access to the database.
Server	Displays the name of the current server but cannot be changed here.
Filename	Displays the name of the file but cannot be changed here.
Database Type	Changes the type of database to a limited number of types. You can change this by dropping down the list if you are manager of the database. The types listed in this field are: Standard, Library, Personal Journal, Address Book[4] or Multi DB Search.[4] You do not need to know any details about these database types for the Application Development I exam.

continues

Table 1.3 Continued

Field	Value and Information
Encryption	Click this button to set local encryption options for this database. Local encryption prevents someone without a valid Notes ID file from reading the database if they are sitting at your computer. This is especially useful for shared machines or laptops. The types of encryption available include: ■ Strong—Slower document access. Recommended if security is the most important consideration. The file cannot be accessed by disk compression utilities. ■ Medium—Most recommended level of encryption because of good security and acceptable speed of access. The file cannot be accessed by disk compression utilities. ■ Simple—Limited security features but fastest document access. File can be accessed by disk compression utilities.
Replication Settings	This button will allow you to set replication options for this database, including the capacity to define what documents, folders, and views will be replicated into and out of this database. Replication is not discussed in depth on the Application Development I exam.
Replication History	This button will allow you to investigate the replications that occurred to and from this database. Also, if you clear the replication history, you can force a complete replication of this database.
Disable Background Agents in This Database	This option limits the use of agents in the database. Agents that run at regular intervals or after a particular event are called background agents. Selecting this tab turns background agents off but allows other types of agents to run.

Field	Value and Information
Allow Use of Stored Forms This Database	Stored forms must be allowed in order to send documents from this database to mail users, so in that they can clearly display the data on the document.

[4]These types are available starting in Lotus Notes Release 4.6.X.

The Information Tab

The next tab on the Database Properties InfoBox is the Information tab, shown in Figure 1.3. This tab shows basic data about the size of the database, including the number of documents in the database and the percentage of the database that is in use. The Compact button on the Information tab eliminates any unused white space in the database, making the database smaller and more efficient. Any time the database percentage used falls below 90 percent, the database should be compacted. The Information tab also displays the database's creation and modification dates. The User Activity button on the tab enables the user to see any reads and writes to the database. Finally, the Information tab shows the Replica ID for the database. A replica ID provides a unique identification number for the database. All replicas of a database will share the same replica ID.

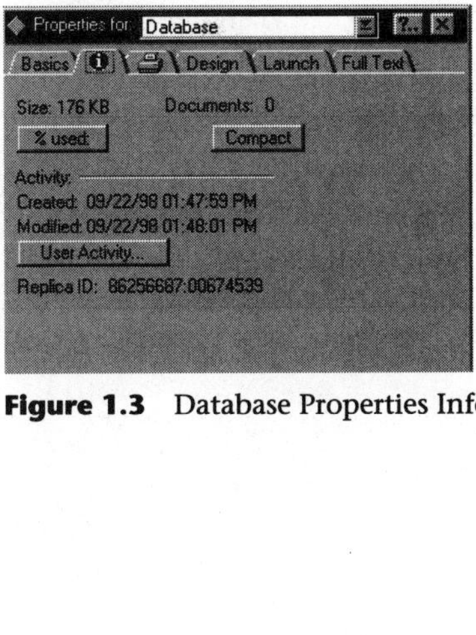

Figure 1.3 Database Properties InfoBox Information tab.

The Print Tab

The Print tab, shown in Figure 1.4, defines the settings to be used for printing the database. It allows you to create headers and footers for the database pages and specify the font, font size and font style. Headers and footers can contain automatic elements such as the time, the date, and the page number. By default, headers and footers are printed on each page except the first. You can, however, force them to print on the first page as well by using the check box at the bottom of the page.

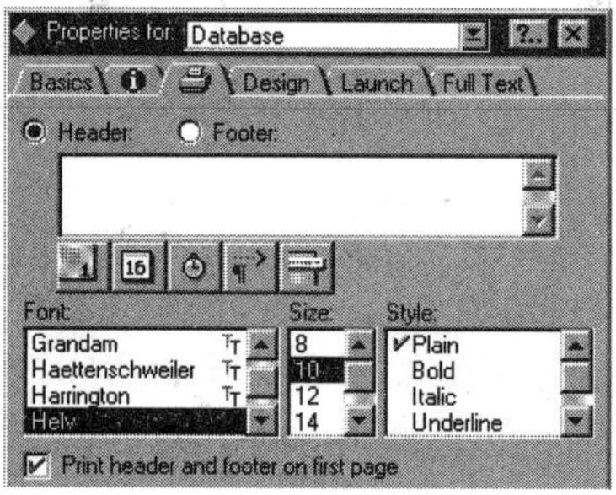

Figure 1.4 Database Properties InfoBox Print tab.

The Design Tab

The Design tab, shown in Figure 1.5, enables you to define properties of the database design, including template and database catalog information. A database can inherit design information from a database template. This can occur at creation or any time after creation. The database can also become a template, allowing other databases to inherit its design elements. The Design tab has the capacity to hide the database from the Database Catalog, though it is shown in the catalog by default. The Database Catalog option also determines whether the database will be listed in the Open Database dialog box. Finally, the Design tab enables you to choose not to mark modified documents as unread. This prevents documents that have been changed by background agents from being marked read before the user has actually opened and read the documents.

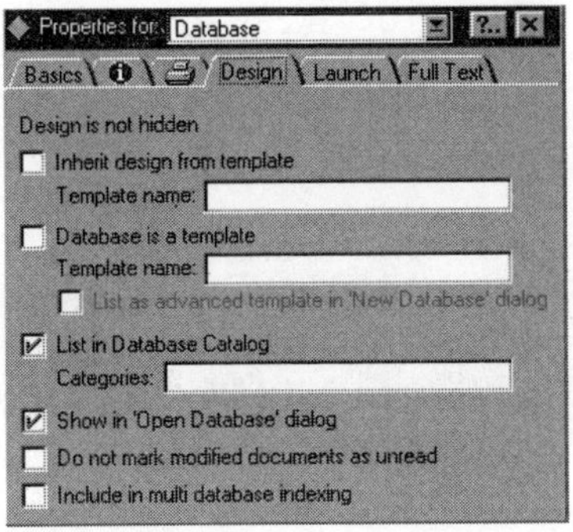

Figure 1.5 Database Properties InfoBox Design tab.

The Launch Tab

The Launch tab, shown in Figure 1.6, sets the properties for how the database appears when it opens.

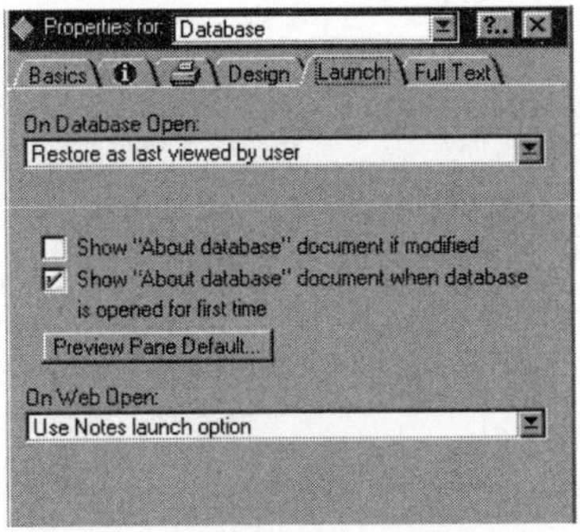

Figure 1.6 Database Properties InfoBox Launch tab.

The On Database Open drop-down list options are:

- `Restore as last viewed by user`. This is the default setting for this option and displays the database in the same place the user last opened it. The About Database document options for this setting can be set to display only the first time the user opened the database or any time the About document changes.
- `Open "About database" document`. This option opens the About Database document every time the user opens the database.
- `Open designated Navigator`. This option launches a specific navigator each time this database is opened. The Navigator is displayed

in the navigator pane, on the left side of the screen. The Navigator is displayed with all views. Navigators are described in depth in Chapter 5. The About Database document options are available in this option.

- `Open designated Navigator in its own window.` This option opens the designated Navigator as a full screen. The Navigator is displayed each time the user exits a document using the Escape key. Again, a specific Navigator must be selected. The About Database document options are not available in this option.
- `Launch 1st attachment in "About database."` This automatically starts the first attachment included in the About Database document. There are no About Database document options available.
- `Launch 1st doclink in "About database."` This automatically starts the first doclink included in the About Database document. There are no About Database document options available.

Navigators and About Database documents will be described in further detail in Chapter 5.

The Full Text Tab

The Full Text tab defines the settings for full text indices available in Notes. A full text index can be created for any database, either locally or on a server, by clicking the Create Index . . . button, as shown in Figure 1.7.

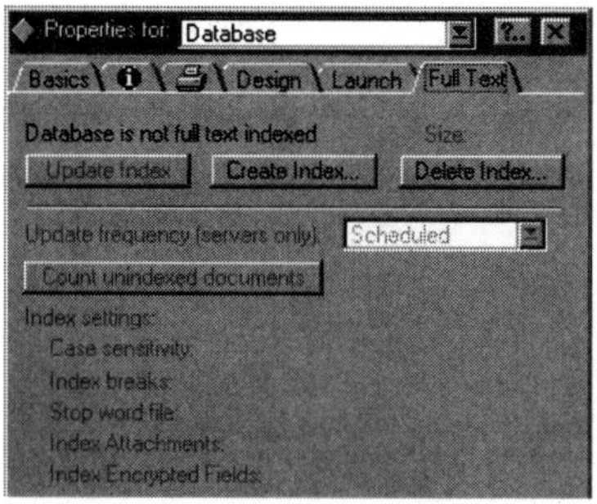

Figure 1.7 Database Properties InfoBox Full Text tab.

The Full Text index allows the user to perform Boolean searches on the database, which will return documents based on the following options:

- `Case-sensitive index`. A case-sensitive index causes words in different cases to be listed separately in the index. Notes and *notes*, for example, each would have separate listings. This will increase the size of your index.
- `Index attachments`. This option indexes any text in attachments that is included in the database. This is especially useful for document library databases, which have many text documents attached to the database. This will increase the size of your index, depending on the size of the attachments.
- `Index encrypted fields`. This option adds the text from any encrypted fields into the index. It will increase the size of your index, depending on the size and number of encrypted fields. Note that the security of encrypted fields is abridged by including the text in the index.
- `Exclude words in stop file`. Certain words, such as *a*, *an* and *the*, are not significant to an index and can be excluded by using this option. The designer can either use the default stop word file, `DEFAULT.STP`, or create one specifically for a database. This option can decrease the size of the index.
- `Index breaks`. The two choices for index breaks are Word Breaks only and Word, Sentence, and Paragraph. The Word Breaks only option indexes all words, regardless of location. If the index is based on Word, Sentence, and Paragraph breaks, however, the designer can index the database based on the location of certain words. A search for *replication sentence troubleshooting*, for example, will find only documents where the words *replication* and *troubleshooting* appear in the same sentence.

NOTE

Notes has two special documents called `About This Database` and `Using This Database`, which contain primary help information for database users. Notes creates both documents by default in every database. You should edit them to include information specific to the database you created. To edit the documents, select the Design . . . Other view from the Navigation pane. The user accesses these documents from the Help menu. When an index is created, these two documents are automatically added to the search index.

On servers, there are several options for updating the index based on the following choices: Immediate, Daily, Scheduled, and Hourly. When running on the server, the Periodic Full Text Indexer updates the index. On a workstation, these options are not available. In File . . . Tools . . . User Preferences on the Basics panel, however, you have the option to `Enable local background indexing`. If this option is enabled, the indexes will be updated upon workstation startup.

EXAM TIP

To pass the Application Development I exam, you must be able to distinguish between various Database Property InfoBoxes for each element of a database. The exam tests your knowledge of the property InfoBoxes with questions that first describe a particular behavior in a database, then require you to determine which box sets or creates that behavior. The Database Properties InfoBox discussed before will draw a good percentage of the questions. InfoBoxes for other elements of the database, such as Form, Document, View, Column and others, will be discussed later in this book.

Sample Database

To pass the Application Development I exam, it is important to work with the software to create an actual application. This allows you to use the various InfoBoxes, formulas and other settings tested on the exam. At the end of each chapter, there are steps to help you create a sample application. This sample application is geared toward specific goals of the Application Development I exam—and thus excludes the graphic and other formatting elements significant in creating an application for real use. Instead, we focus on the direct tasks covered on the exam. Included in this section are screen captures that show what your application might look like. Keep in mind that often, there is more than one right way to do things in developing a database. This section gives you some ideas for practice.

The sample database we will create is a discussion-type database used for Question tracking. It could be used by a help desk, for example, to assist users who would rather type in a question and have someone respond than wait on hold for phone support. Think about what makes this database appropriate for a Lotus Notes database. First, if the users already are familiar with Lotus Notes, they do not need to open a special application and can easily create questions. Second, by using the database rather than e-mail, the answers to users' questions are immediately available for all users who have database access. This saves the help desk personnel from answering the same question numerous times, and it helps users who are too shy or too busy to ask the question themselves. Our database will be creatively titled `User Question` Database.

To begin, first create a new database according to the following steps:

1. Choose File . . . Database . . . New from the menu, or press Ctrl+N to create a new database.
2. This displays the New Database dialog box. Choose `Local Server`, because we will create the database on the local machine. Make the database title `User Question` Database, and make the filename `USERQUE.NSF`. The filename can be longer in some file systems, but I chose a name that follows the 8.3 file naming convention for compatibility purposes. The database will be created in the local `\NOTES\DATA` directory by default.
3. Accept the default size limit of 1GB, and do not apply any local encryption.
4. The database will not be based on any template, because we want to create our forms and views from scratch (for practice). The completed New Database dialog box is displayed in Figure 1.8.
5. Click OK to create the database. After the database icon is on your workspace, you can right-click to bring up the Database Properties InfoBox.

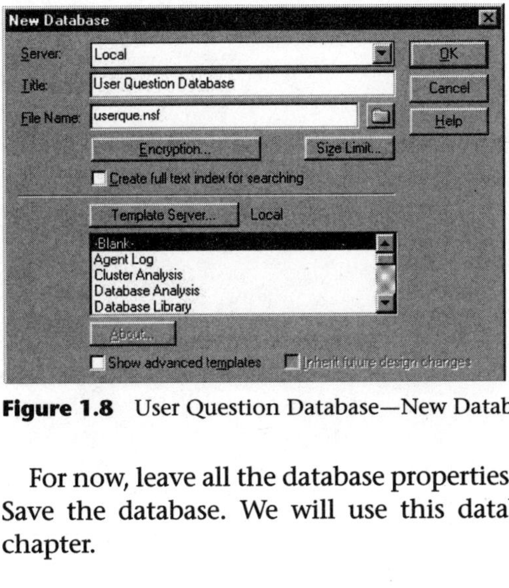

Figure 1.8 User Question Database—New Database dialog box.

For now, leave all the database properties at their default settings. Save the database. We will use this database again in the next chapter.

Review Questions

1. What type of database should Austin create if he wants to create a newsletter for his company that automatically updates users with daily news release information?

   a. Discussion

   b. Tracking

   c. Reference

   d. Broadcast

2. Would Tyler recommend a Notes database to his clients if they need an application that performs advanced mathematical equations and ad-hoc queries?

   a. Yes

   b. No

3. Maggie is a new Notes developer. She is developing her first database and wants to know whether to create it on her local machine or on her server. When she calls her administrator, where should the administrator tell her to create it?

   a. Local Machine

   b. Server

4. The Notes database default size limit is:

   a. 2GB

   b. 4GB

   c. 1GB

   d. 500MB

5. The maximum Notes database size is:

   a. 2GB

   b. 4GB

   c. 1GB

   d. 500MB

6. What method(s) can Emily use to create a new Notes database?

   a. Copy another Notes database and change any necessary design elements.

   b. Convert a Microsoft Word document into a Notes database.

   c. Base a database on one of the design templates included with Notes.

   d. Create a new, blank database and add all the necessary design elements from scratch.

7. The following file extensions are allowed for a Notes database:

   a. .NSF

   b. .NS3

   c. .NDB

   d. .NDX

8. After he created a new database, Doug realized he only allowed it to be a maximum of 2GB. The requirements for the database said it needed to be able to grow to 5GB. What should he do to correct his mistake?

   a. Nothing. Notes database sizes cannot be changed after they are created.

   b. Nothing. Notes databases cannot be 5GB.

   c. Change the `Size Limit` option on the Database Properties InfoBox Design Tab.

   d. Make a new replica of the database and change the Size Limit option to 5GB while creating the new replica.

9. C.J. creates a new database and tries to name it *The Scottish Historical Cookbook and Recipe Database.* What should you tell him when he calls the help desk and complains that he cannot type in the complete name?

   a. A maximum of 32 characters is allowed in a database title. Limit the name length.

   b. For a longer title, you have to type the name in a text document and use copy and paste to fit it in the field.

   c. Click the Size Limit button to allow a longer limit on the title.

   d. There are no spaces allowed in the Database title. If you leave out the spaces, the title will be accepted.

10. Databases that are being used to mail documents to users need to have this option enabled, to allow users to view the documents correctly when they receive them:

    a. Disable background agents

    b. Allow use of stored forms

    c. Inherit design from the StdR4Mail template

    d. Make Database Mail enabled

11. Samantha wants to have a graphical Navigator displayed to her database users. She wants this Navigator displayed in its own window. Where does she find the option to enable the Navigator?

    a. New Database dialog box, Advanced Options button

    b. Database Properties InfoBox, Design tab

    c. Database Properties InfoBox, Launch tab

    d. View . . . Show . . . Navigators . . . Options

12. When Elizabeth creates a new database, she wants to be sure that users can search all elements in the database. She creates a full text index and uses options to allow searches on the following items:

    a. Text attachments
    b. Encrypted fields
    c. Graphic attachments
    d. About This Database and Using This Database Documents

13. The following are characteristics of a Notes database template:

    a. Uses the extension .NTF (Notes Template Facility)
    b. Contains a full set of documents, to give examples of how the database should be used
    c. Contains all the necessary design elements, including views, form and agents
    d. Must be created from scratch

14. Ashley is filling a database request for a database that will replace the current business card and stationery ordering process. The process currently involves three paper forms, which are sent to five people for approval and processing. Which database type is Ashley likely to choose?

    a. Broadcast
    b. Tracking
    c. Reference
    d. Workflow

15. Jack is defining the options for triggering an event when the database he is creating opens. Which events can he trigger at the opening of the database?

    a. Open first document in default view.
    b. Launch first attachment in the database.
    c. Open designated Navigator.
    d. Launch first doclink in the About document.

16. Rachel wants to create a database that can only be opened by users who know about it. This means she does not want it available when users select `File . . . Database . . . Open`. How can she set this option?

    a. She can set the `Hide From Open Database` option in the New Database dialog box.

    b. She can set the `Hide From Users` list option in the Database Properties InfoBox, Design tab.

    c. She cannot set an option to hide the database from the File . . . Database . . . Open dialog box.

    d. She can deselect the List in Database Catalog option on the Database Properties InfoBox, Design tab.

17. Alex is creating a new database, and Taylor tells him to create it on his local workstation instead of on the server. What reason does Taylor give Alex for this?

    a. Users may start using the database before it is complete.

    b. The database may corrupt other databases when new forms and views are created in it.

    c. Other Notes databases and applications may try to use the data in the new database.

    d. Design functions will not work on the server, only on the workstation.

Answers

1. d. Broadcast. A broadcast database allows the designer to forward documents.
2. b. No
3. a. Local Machine
4. c. 1GB
5. b. 4GB
6. a, c, and d. Copy from another Notes database, create from a template and create from scratch.
7. a and b. .NSF and .NS3
8. b. Notes databases cannot be 5GB.
9. a. The database title has a maximum of 32 characters.
10. b. Allow use of stored forms.
11. c. Database Properties InfoBox, Launch tab.
12. a, b, and d. Text attachments, encrypted fields and About This Database and Using This Database documents.
13. a and c. Uses the extension .NTF and contains all of the necessary design elements.
14. d. Workflow
15. c and d. Open designated Navigator and Launch first doclink in the About document.
16. d. Deselect the List in Database Catalog option.
17. a. Users may start using the database before it is complete.

CHAPTER 2

Creating the Form

Chapter Objectives:

After reading this chapter, you should be able to answer questions based on the following objectives:

- Create a new form.
- Name a form, including the use of Synonyms and Cascading Names.
- Force documents created with a form to start in Edit Mode when opened.
- Enable a form to inherit data from another form.
- Enable a form to be completed anonymously.
- Create a form that will merge replication conflicts.
- Create a form that will be stored with the documents that were created with that form.
- Create a form that uses version control.
- Add static text to a form.
- Organize a form using sections.
- Use hotspots and other objects on a form.
- Limiting viewed information on a form using Hide-When options.
- Creating and using a subform.

The most basic element in a database is the form, which is similar to a stencil or template for the data you want to gather. The form gives structure to the data, so users can input and view the data. When a user inputs data onto a form and saves it, the user creates a document. The exam concentrates on using Form Properties to create desired effects on a form and on creating basic elements of the form. This chapter describes how to create a new form and discusses all properties available for the form. After the form is created, the chapter will describe elements such as static text, graphics, sections, and others that will be on the Application Development I exam.

To create a new form, open the database and choose Create . . . Design . . . Form. You can also go to the Design . . . Forms view from the Navigation pane and click the Create Form SmartIcon.

NOTE
Different SmartIcons are displayed depending on the context. The Create Form SmartIcon only will be available when you are in the correct view.

The design view will open to the new form, displaying the Form Builder window in the top pane and the Design Pane in the bottom, as shown in Figure 2.1. The form builder window enables you to add elements such as static text, graphics, and fields. The design pane enables you to create formulas for the form. You can hide the design pane by toggling it on and off, using the View . . . Design pane. You may also want to show the ruler to aid in layout, by choosing View . . . Ruler or by clicking the View Ruler SmartIcon.

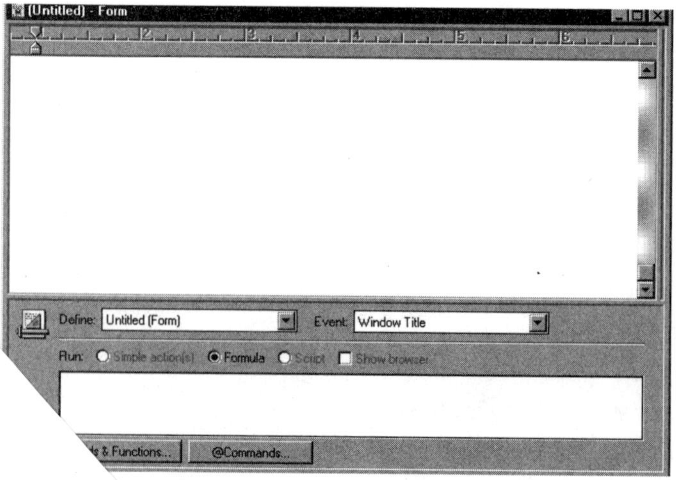

Form Builder window.

Form Properties InfoBox

After you create the form, you will want to set the appropriate properties using the Form Properties InfoBox. To open the Form Properties InfoBox, do one of the following:

1. Right-click on the form and choose Form Properties from the shortcut menu.
2. Choose File . . . Document Properties while in design mode on a form. This option will display the Form Properties in this context.
3. Click the Design Form Properties SmartIcon.

The Form Properties InfoBox is displayed. The options in this infobox are tested in depth on the exam. We will examine the options on each tabbed page in the following paragraphs.

The Basics Tab

The first tab on the Form Properties InfoBox is the Basics tab, as displayed in Figure 2.2. The Basics tab enables you to set main options, like the name and type of the form. The following paragraphs describe each of the options available.

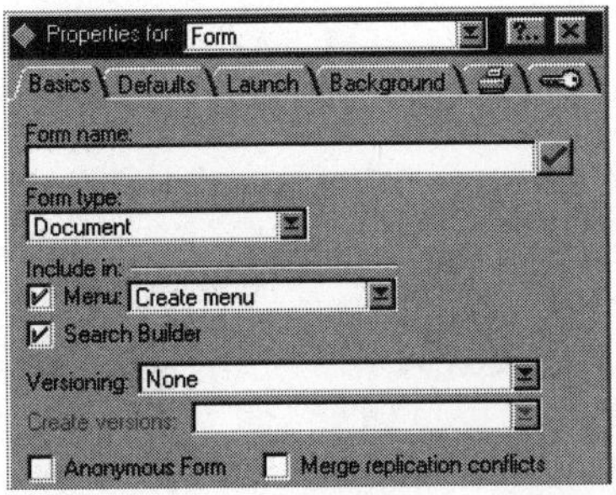

Figure 2.2 Form Properties InfoBox, Basics tab.

- *Form Name.* The form name should be descriptive, including information about the form's purpose or the type of data on the form. The form name can be up to 256 characters, including alpha-numeric characters, spaces, and punctuation. Only the first 64 characters of a form name, however, will be displayed. The form name is case sensitive. Form names also can use *synonyms,* which are the names used internally by Notes. Synonyms can be useful in situations where the name is long enough to be cumbersome when using it in formulas, or in situations when you want the form to have multiple names, such as during the process of translating the application to another language. To create a synonym for a form name, type in the form name first (this name will be used in the Create menu). Then use the | symbol (pipe or vertical bar). Next, type in the synonym, which will be used internally and in formulas—and will also be used in other Notes design elements, including views and keyword choices. For example, Figure 2.3 shows the Book order form, with the synonym name of Order1. After you have typed in the form name and the synonym, click on the green box to have Notes check the syntax.

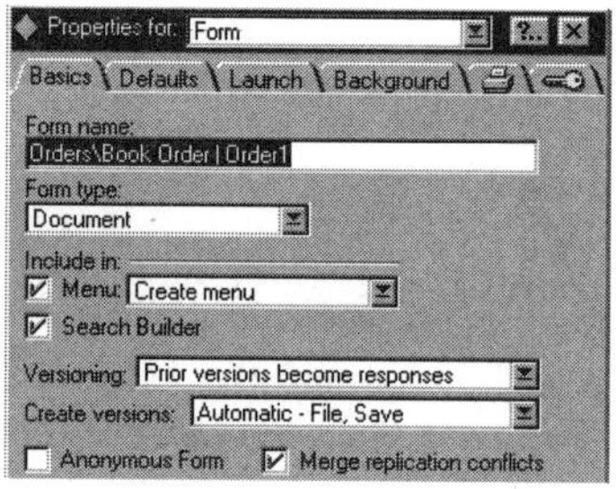

Figure 2.3 Book order form.

NOTE

Form names can also be cascaded off the Create menu. In some cases, you may want to group a set of forms together under a menu choice. To cascade a form name, first type the name of the menu under which you want to cascade the name. Then type a \ (backslash) and the name of the form. Be aware that the menu choice portion of the form name has to be precisely the same, including capitalization, on all forms you want cascaded. It is usually easiest to copy the first portion of the name from a previously created form when you are placing multiple forms under the same cascaded menu choice. Results of this option are displayed in Figure 2.4.

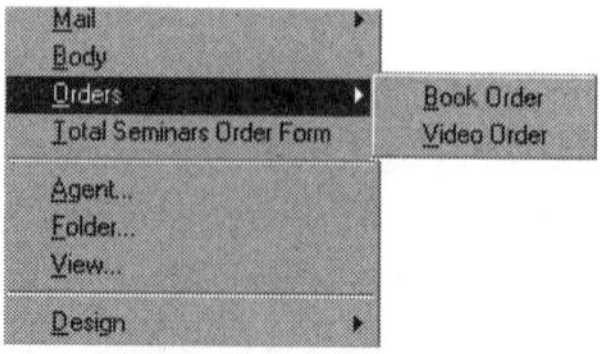

Figure 2.4 Cascaded form names.

Three types of forms are available in Notes: *Document, Response,* and *Response-to-Response.*

1. *Document* is the default form type. It should be used unless a hierarchical relationship between documents is necessary. When a document form type is used, the documents created by that form are known as *main* or *parent* documents. The terms are used interchangeably in document types.
2. *Response* documents will always be associated with a main document. The response document will be a child document to a main document and will usually inherit some information or data from this parent document. For example, Cassandra created a new question document in the Technical Support database, and Mike wants to respond to the question. When he creates a response document, he will want some of the information from Cassandra's document, such as the date and subject, to be included in his response. This process, inheritance, is discussed more thoroughly later in the chapter.
3. *Response-to-Response* documents can be associated with either a main document or a response document. This feature enables a third hierarchical response level in the database. In the previous scenario, for example, Scott, the database moderator, needs a third type of document to comment on either the questions or the answers. This document needs to be a response-to-response form type.

NOTE

When discussing response and response-to-response forms, you also need to understand the difference between children and descendants. Consider a main document to be a parent. Then consider a response document that is directly responding to that main document to be a child. Next, consider a response-to-response document that responds to the main document also to be a child. Finally, consider a response-to-response document that responds to a response document as descendant, not as a child. You will need to know this information when selecting documents for a view. Figure 2.5 shows the possible relationships between forms.

- *Include In.* The form can be included in the Create Menu, the Create . . . Other dialog box, or neither. If the designer chooses to place the form in the Other . . . dialog box, a separate option on the Create menu will be displayed called Other . . . . This option then displays a separate dialog box from which these forms can be selected. This option enables the designer to keep the Create menu less cluttered. The form name, not the synonym, will appear in the menu options or in the Other . . . dialog box. By default, the form is included in the Search Builder to enable users to search for information by form. If you are securing or hiding the form, do not include the form in the Search Builder.

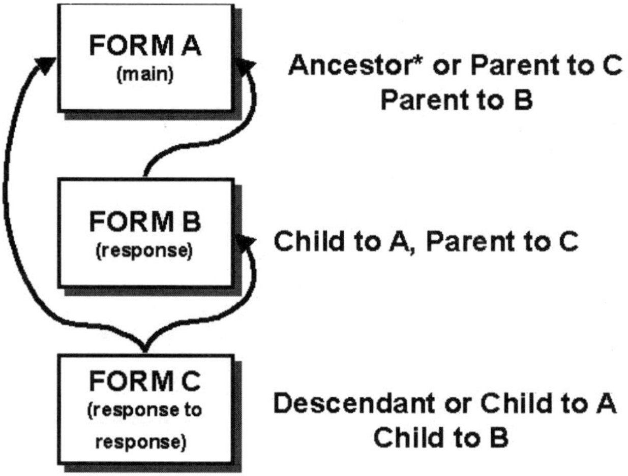

Figure 2.5 Document relationships.

NOTE
The Search Builder helps designers choose documents for views and agents. This concept is discussed fully in later chapters.

- *Versioning.* Version tracking enables an original document and any edited versions of the document to be saved. Version tracking can be especially useful when each change to a document needs to be maintained and visible. There are four version tracking options:

1. *None.* This option is the default option. There will be no version tracking. When a document is edited and saved, it replaces the original document.
2. *New Versions become responses.* In this option, when a document is edited and saved, it will become a response to the original document. This option can help prevent replication and save conflicts. When replication occurs—if users on multiple servers have

edited the document—there only will be multiple response documents, rather than conflicts. This option is good if the original information is more important than the changes being made. If multiple people were working on the chairman's major speech, for example, the basic text would remain intact while each writer put his or her spin on it. The chairman and the writers could view the different versions and compare them with the original.

3. *Prior versions become responses.* When this option is enabled, the most recently saved version of a document becomes the main document. All predecessors are saved as response documents to the main document. The original document will be the first response document. This option should be used when the focus is on the changes, rather than on the original.
4. *New versions become siblings.* With this setting, all new versions of a document become other documents at the same level, much like additional main documents. This feature is used if the time stamp of the revisions is unimportant. It does make it more difficult to visualize the relationship between documents, however, because they are all at the same level in the view.

When version tracking is enabled, an option also exists for Create Versions, which has the following features:

- *Automatic—File Save.* When a document is edited and saved, it will automatically use the version tracking setting to create a new document.
- *Manual—File, New Version.* In this option, the original document will be replaced by the edited version, unless the user chooses File . . . Save as New Version at save time.
- *Anonymous Form.* Notes R4.X enables documents to be created anonymously. This feature is often used for surveys or ballot box-type applications. The automatic field $UpdatedBy is replaced with an $Anonymous field with a value of one. If you want to create a form that enables users to respond without being identified, even by developers or administrators, mark the form anonymous.
- *Merge Replication Conflicts.* This field can prevent replication conflicts that occur when the same document is edited simultaneously on two different servers. When replication occurs between two servers—where the same document has been edited on both

servers—the documents will create a replication conflict. In Notes R4.X, however, replication works at the field level, rather than at the document level—which means that when changes are detected on a document, only the changed fields are replicated —rather than the entire document. Also, if Merge Replication conflicts is enabled, and two different fields have been edited, there will be no replication conflicts. Mike and Brian, for example, are both working on the same set of documents. The documents exist in two replicas of a Notes database on two servers. They both have editor-level access to the set of documents, and they both edit those documents to add to the documentation. They both edit the same document, but Mike edits the Topic field and Brian edits the Body field. If Merge Replication conflicts is enabled, when replication occurs the two fields on the document will be updated, and no replication conflict will be created. If both Mike and Brian update the Topic field, however, a replication conflict will be created.

NOTE

A Save Conflict occurs when the same document is opened and edited simultaneously by two users on the same server. In the previous example, if Mike and Brian were both working on the same database on the same server, any time they both edited the same document simultaneously, a save conflict would be created.

The Defaults Tab

The next tab on the Form Properties InfoBox is designed for default settings and is called Defaults. The settings for this tab, as displayed in Figure 2.6, are discussed in the following paragraphs.

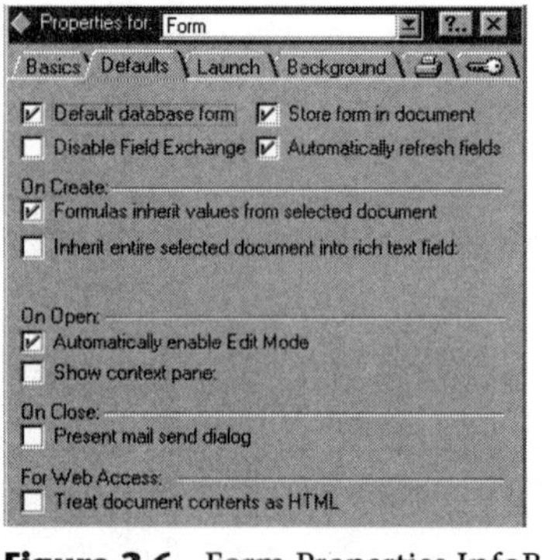

Figure 2.6 Form Properties InfoBox, Defaults tab.

- *Default Database Form.* This setting makes the selected form the default for the database and means that if the form associated with a particular document cannot be found, this form will be used to display the document. Only one default form per database can exist, and it is distinguished in the Design . . . Forms view with an asterisk.
- *Disable Field Exchange.* Field exchange (Notes/FX) enables Lotus Notes applications and OLE server applications to share or exchange field information. This option enables Field Exchange to be disabled. The fields, however, can remain on the form. Notes/FX and Field Exchange is only covered on the exam as far as describing how to disable it. The Lotus Notes Help database is available to give you more information on Notes/FX.
- *Store Form in Document.* By default in Notes, the form and data are stored separately. By selecting this option, however, the data and form will be stored together—which is required for documents or forms that will be e-mailed to users or for mail-in databases. The documents will be larger, however, and previously created documents will not reflect any design changes made to the form.

NOTE

The Allow the Use of Stored Forms in This Database option on the Database Properties InfoBox must be enabled to use this option.

- *Automatically Refresh Fields.* Any time it is necessary for users to see updated information on a form—such as calculations on an order form— as they complete it, this option should be selected. It will force every field on the form to be recalculated each time the user moves to a different field. Be aware, however, that this option will slow movement through the form, especially on a form with multiple calculated fields.

NOTE

The phrases *On Create* and *On Open* can be confusing, but the distinction is important. *On Create* refers to the time when the user completes a form, before it has been saved. *On Open* refers to any time the document is opened after it has been saved. Therefore, when a user creates a question in the database by choosing Create . . . Question from the menu, the document goes into create mode. After the user saves the question, the help desk person then opens the question document.

- *On Create.* Two options for the form exist at creation time. Both of these options relate to the idea of inheritance. *Inheritance* is the capacity of a form to pull some information for its fields from another form. This feature only exists when composing documents. After the document has been created and saved, it can access information only on another document through formulas or LotusScript, not through inheritance. Additionally, for inheritance to work correctly, the user must have a particular document selected in a view when they create the new document that needs to inherit information. This document can be any form type (main document, response, or response-to-response). The two On Create: Inheritance options are discussed in the following paragraphs.

- *Formulas Inherit Values from Selected Documents.* When this option is selected, fields can inherit their values from fields on the selected document. This helps avoid unnecessary typing and mistakes when transferring information from one form to another. To set up inheritance, set the default value for the field to the name of the field from which it should inherit its value. For example, the designer of our call tracking database wants the call response to include the question that the user asks. The question is typed into a field called Question. In the call response form, therefore, there should be a field for which the default value is *Question,* and the inheritance option should be enabled.
- *Inherit Entire Selected Document into Rich Text Field.* This option enables an entire document to be copied into a rich text field on the new document. It can be copied into the rich text field in one of the following ways: as a link, as collapsible rich text, or simply as rich text. Figure 2.7 displays additional options when this selection is enabled. Note that a rich text field must exist on the form which you are designing to have this option. The link option will create a doclink to the selected document. The collapsible rich text option will create a collapsible section that has the contents of the selected document. The rich text option copies the contents into the indicated rich text field. An example of this option in use is forwarding e-mail, or sending replies to e-mail that include the history.

To enable inheritance completely, you must use fields whose values are defined by the fields on the selected documents. This portion of inheritance will be described in Chapter 3.

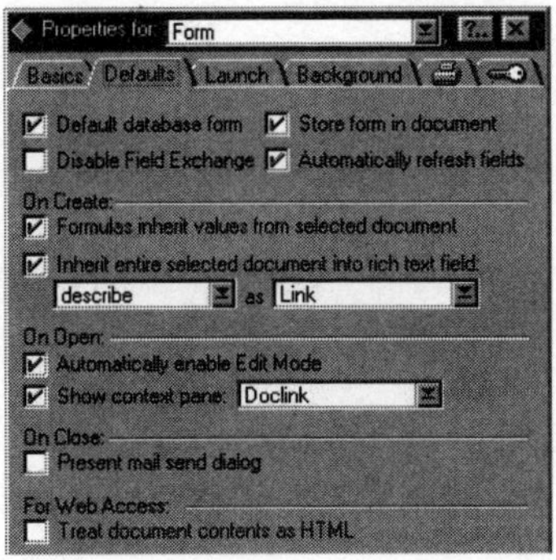

Figure 2.7 Form Properties Defaults tab, on Create Inheritance options.

- *On Open.* Two options exist for opening documents created with this form. These options are described in the following paragraphs:

1. *Automatically Enable Edit Mode.* When this option is selected, each time a document created using this form is opened, it will automatically open for editing. The default for opening a document would ordinarily be read mode. A user would then have to click an action button or use a menu choice to enter edit mode. This option can save time for data entry, when documents will be edited each time they are opened.
2. *Show Context Pane.* A context pane is another partition to the open window that displays additional information to the user. In this option, you can open a context pane to display either the parent document of the current document or a doclink included in the document, as shown in Figures 2.8 and 2.9. This option is especially useful when combined with inheritance options or when used for response documents.

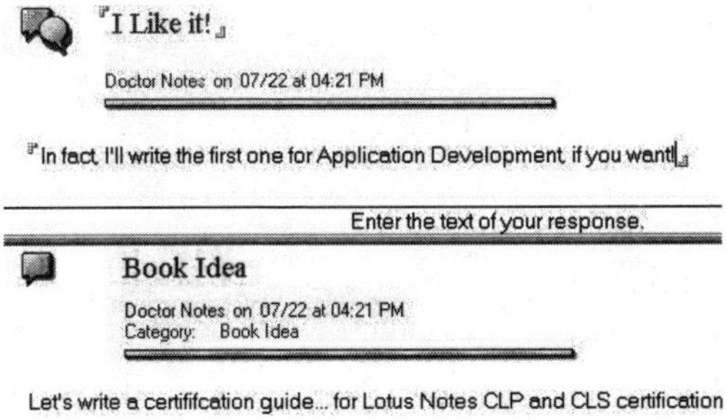

Figure 2.8 Document with parent document displayed in the context pane.

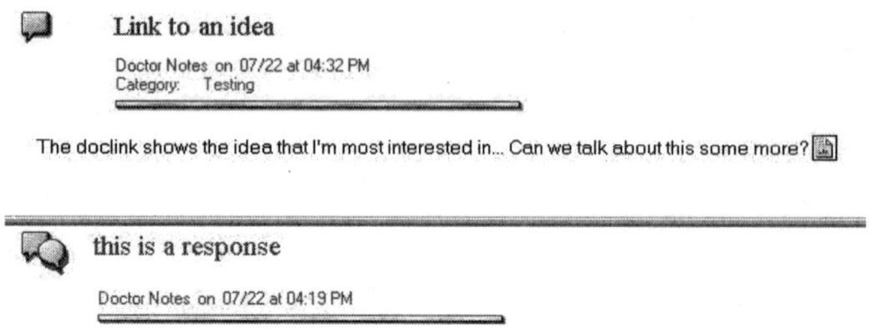

Figure 2.9 Document with context pane displayed for a doclink.

On Close. The designer can specify here that the Mail Send dialog box will appear when a user closes a document. This feature will enable the user to send the document to an e-mail user, save the document, or discard changes to the document. For the document to be e-mailed correctly, the form must have the required e-mail fields for addressing. This topic is discussed in detail in Chapter 6.

For Web Access (4.6.1X). This option will automatically treat the document contents as *Hypertext Markup Language* (HTML) when selected. This option is most useful when the database is being developed primarily for World Wide Web-based access. This option will not be tested on the Application Development I exam.

The Launch Tab

The Launch tab on the Form Properties InfoBox, as shown in Figure 2.10, enables a doclink, attachment, or other object on a form to be launched automatically. This feature enables the designer to place information from other documents, databases, or applications on the forms they create. The default setting for the option –None– ensures that nothing is launched automatically. Links or objects can still be launched manually by the user.

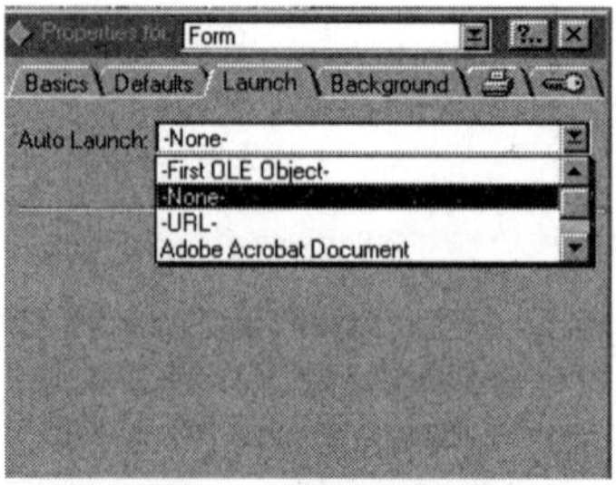

Figure 2.10 The Launch tab.

The drop-down list contains options to automatically launch other objects. The options include:

- First attachment
- First document link
- First OLE object
- *Uniform Resource Locator* (URL)
- Any object from the OLE-enabled applications installed on the workstation

NOTE
Users will see different lists for every workstation, because each workstation will have different applications. Notes lists all the OLE-enabled applications in this option, because these applications can launch any objects or attachments created with them.

Whether the designer chooses to launch the first attachment, document link or URL, the object will be launched immediately when the user selects Create . . . *Formname*. If the designer chooses any of the OLE options, the following additional options are available, as shown in Figure 2.11.

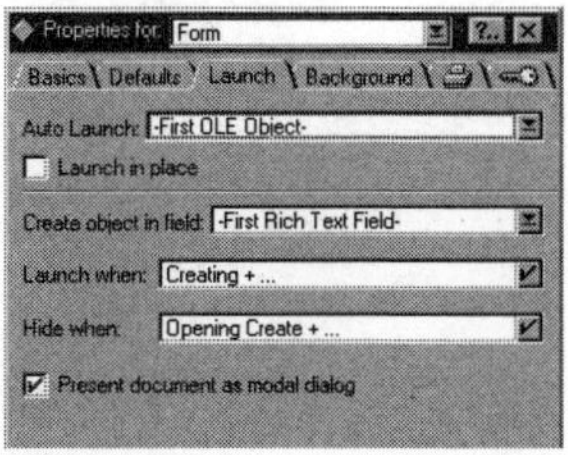

Figure 2.11 Additional Launch options.

- *Launch in place.* OLE objects can be launched either in place or out of place. When objects are launched out of place—the default setting—the new application launches separately and becomes the active window. When an OLE object launches in place, the application is opened within a Notes window.
- *Create object in field.* When creating a new OLE object, the designer chooses whether it is created in the first rich text field or

in any other rich text field on the form. If one already exists on the form, it will be listed here. If the object should not be displayed as part of the form, the designer can also select -None-.

- *Launch when.* This option enables the designer to toggle on and off the three options of Creating, Editing, and Reading, all of which are toggled on by default. The Launch when option must be set with the purpose of both the object and the form in mind. For example, if the object should be created when the document is created, Launch when creating must be selected. If the document automatically launches in Edit mode (an option on the Defaults tab), Launch when editing should be selected.
- *Hide-when.* This option enables the designer to hide the Notes document automatically upon certain events, which means that the user will only see the application that launches the external object, and they will never see the Notes document. The Notes document can remain hidden during the following events: Opening Create, Opening Edit, Opening Read, Closing Create, Closing Edit, and Closing Read. All the Opening options are selected by default. Each of the Closing options is available for selection only when the corresponding Opening option is already selected.
- *Present document as modal dialog.* This option enables the designer to launch the document or object into a dialog box without enabling the end user to access the Notes menus while the object is active.

The Launch options on the Form Properties InfoBox will not be tested in depth on the Application Development I exam. This information is still necessary, however, for questions or for a basic understanding of building a form.

The Background Tab

Available in Lotus Notes R4.5 and higher, some of the Background tab options can be found on the Basics tab in earlier versions. The Background tab sets the background color and also any background graphics you want displayed on the form, as shown in Figure 2.12. The fields on the Background tab are discussed in the next few paragraphs. Be aware that only the most basic information about creating a background is tested on the exam.

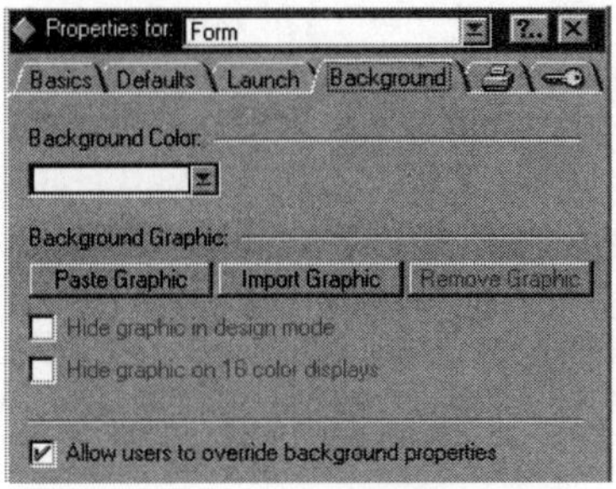

Figure 2.12 The Background tab.

- *Color.* This drop-down box will let you choose a background color for the form. Keep in mind that everyone has different display capacities on monitors, ranging from black-and-white to 16 million-color monitors with advanced graphics capabilities. You will want to use colors that can be viewed on most machines in your environment.
- *Background Graphic.* You can use a bitmap image copied from another application as the background for your form by choosing the Paste graphic option. The graphic already should be on the clipboard to use this button. In the more recent versions of Notes, you also have the ability to import graphics of various formats by clicking the Import graphic button. This concept will not be tested on the exam.
- *Hide Graphic in Design Mode.* Sometimes it is easier for the developer to design the fields and formulas on a form without viewing the graphics. This option enables graphics to be hidden during design time.
- *Hide Graphic on 16-color Displays.* This option enables the graphic to be hidden from users with lower-quality displays.
- Allow users to override background properties. This option allows users to override the colors or graphics chosen by the designer with their own selections. Enable this option to give users more flexibility or to compensate for when some users' monitors and video cards may be of lower quality.

The Print Tab

The Print tab is similar in most of the Properties infoboxes available in Notes, including the Database Properties InfoBox and the

Document Properties InfoBox, as shown in Figure 2.13. The Print tab defines the settings to be used for printing the documents created with this form. It will allow you to create headers and footers for the pages in the database and specify the font, font size, and font style. Headers and footers can contain automatic elements such as time, date, and page number. By default, headers and footers will be printed on every page except the first. You can force them to print on the first page, though, by using the check box at the bottom of the page.

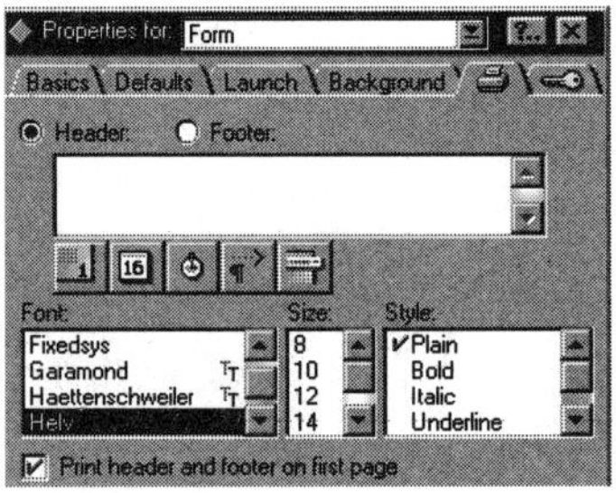

Figure 2.13 The Print tab.

NOTE

The options set here become the default settings for any documents created with this form. They can, however, be overridden by the user with the Document Properties InfoBox for the specific document the user prints.

The Security Tab

Most of the security settings are not tested on the Application Development I exam. Take note, however, of the option Disable Printing/Forwarding/Copying to Clipboard. This setting enables the designer to enforce any security they have placed on the form even outside the Notes application, as it cannot be moved outside the application. This tab is shown in Figure 2.14. For other information about securing Notes applications, please see Chapter 11.

EXAM TIP

The properties described earlier will be tested on the exam in various formats. You may be asked to use this information to troubleshoot form problems or to answer scenario questions. Note that in some cases, additional information was provided for Releases 4.5 and 4.6. Questions on the exam should only come from Release 4.0. Extra information was provided to avoid confusing those of you who are preparing for the exam by using more recent versions of Notes.

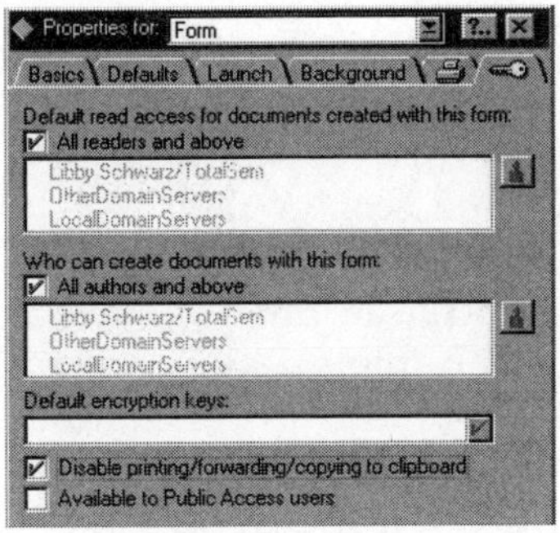

Figure 2.14 Form Properties InfoBox, Security tab.

Form Elements: Static Text

The first element a developer places on a form is static text. Static text gives the form shape and style. Each field usually will be labeled with static text to give information to the user about what to put in the field. Static text can be placed anywhere on the form by clicking and typing. Static text will be displayed on all documents created that use the form, and the text cannot be changed by the user. It is usually recommended to put all the static text on a form first, to create a template for the fields and other design elements you will be adding to the form. Use the normal formatting tools that are available to format this text, including the Text Properties InfoBox as shown in Figure 2.15 and the Text menu choices. The specific tabs on the Text Properties InfoBox are end user topics and are not covered on the Application Development I exam.

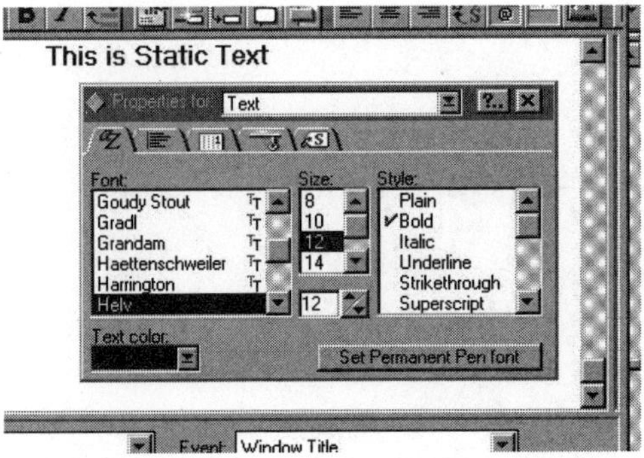

Figure 2.15 Text Properties InfoBox.

Form Elements: Sections

One way to help organize the static text and fields on a form is to organize them into sections. A *section* is a group of paragraphs that can be collapsed into a single line. The section then can be expanded or collapsed, based on user input or designer choice. Sections are useful because a designer can make forms as long as needed by adding more design elements. The more design elements added to the form, however, the more a user will have to continue scrolling through the form. Users often find scrolling down especially long forms tiresome and frustrating. To make the forms more usable and to keep the form shorter, use collapsible sections. Also, sections provide a way for the designer to collapse information that is not necessary or interesting to all users. Sections can be of two types: Standard and Controlled Access. These types of sections are described in the following paragraphs.

Standard Sections

To create a standard section, choose Create . . . Section . . . Standard. It is easiest to create sections after the text and fields for that section have already been created, because sections are more difficult to lengthen after they are created. After a section is created, the first line of text automatically will show up as the section title, and the section will be collapsed and displayed with a small triangle (called a *twistie*) to the left of the section. To customize the section, right-click the section and choose Section Properties, or choose Section . . . Section Properties from the menus, available when the cursor is in the section. The Section Properties InfoBox, shown in Figure 2.16, enables you to change the look and feel of the section. The properties for the section are discussed in the following paragraphs.

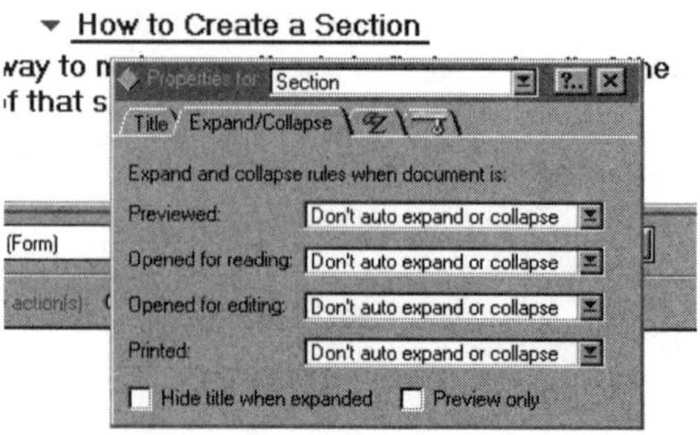

Figure 2.16 Standard Section Properties InfoBox.

- *Title tab.* On the Title tab, you can type in the title you would like displayed at the top of the section. By choosing the Formula radio button, you can use a formula to create the section title. The drop-down boxes at the bottom of the Title tab will enable you to create a border around the section and is set to -None- by default.
- *Expand/Collapse.* The Expand/Collapse tab enables you to set defaults for the display settings of the section. By default, each of the rules on the Expand/Collapse tab are set to *Don't auto expand or collapse,* which means that the section will remain in whatever state it was in when viewed last. Other options are Auto-Expand or Auto-Collapse. You have the ability to set the Expand/Collapse options when the document is previewed, opened for reading, opened for editing and/or printed—which means you can select the option to automatically expand a particular section when you are printing, or automatically collapse a section when reading. This option enables you to use the sections most effectively for organization and form-length purposes. The other options on this tab allow for the title to be hidden when the section is expanded and for the section only to be visible only when the document is in Preview mode.
- *Font tab.* The Font tab enables you to set the font, size, style, and color of the section title.
- *Hide-when.* This tab enables you to set options for hiding the section, either by formula or based on the state of the document. Hide-when options are discussed in depth later in this chapter.

Controlled Access Sections

A controlled access section functions much like a standard section, in that it organizes the text and other elements on a form and limits the length of the form. It has the additional function, however, of limiting who can edit the fields in a section. The additional tabs in the InfoBox for controlled access sections are displayed in Figure 2.17. The *Editors* and *Non-Editors* tabs on the InfoBox set defaults for how the section is displayed for those groups of users. The *Formula* tab defines which users are considered editors. The formula can be Editable, Computed for Display, or Computed when Composed and must resolve to a name or a list of names. The *Title* and *Hide-when* tabs show the same basic options as they do in the Standard Section InfoBox. A few options, however, such as Preview only, have been removed in controlled access sections. It is important to note that anyone with reader access to the database can read the information in a controlled access section unless you use Hide-when options (described later in this chapter). In addition, the controlled access formula will not override the access control list for the database. Controlled access sections are tested in the Application Development II exam and are described in more detail in Chapter 7.

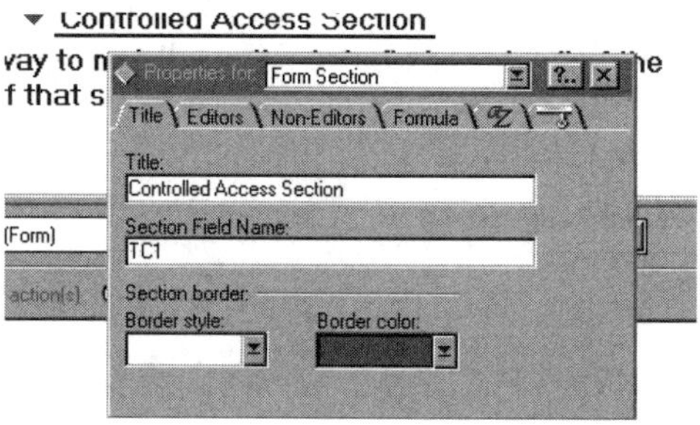

Figure 2.17 Controlled Access section InfoBox.

Form Elements: Hide-When

Hide-when formulas and options enable design elements to be hidden based on a formula or document event. The Hide tab of the Text Properties InfoBox, for example, shown in Figure 2.18, shows how the developer can select events or write a formula to hide the selected design element.

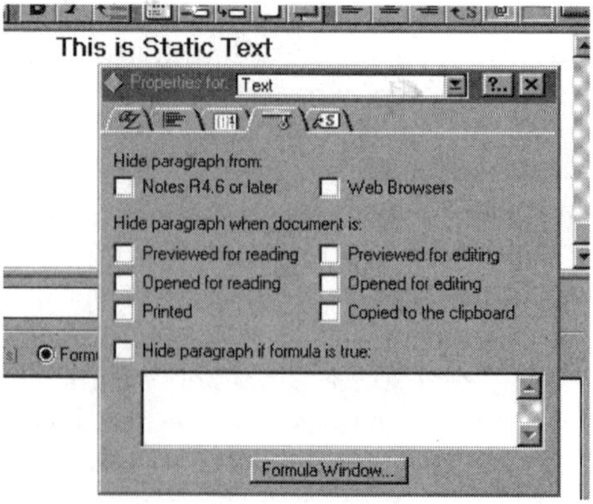

Figure 2.18 Hide-when options.

Document events include the following: Previewed for Reading, Previewed for Editing, Opened for Reading, Opened for Editing, Printed, and Copied to the Clipboard. In addition, a formula based on the current user or events can be added to customize the form further. Hide-when is available for text, fields, sections and other design elements. Hide-when is often used when the designer wants the form to look one way when information is added and another way when data is read. For example, a data entry person may need instructions on the fields that someone just reading the data would not need to see. Hide-when options can be used for the following purposes:

- Displaying one field to enter or edit a value and one field to display the value.
- Showing Actions at some times and not others.
- Displaying particular sections or data only to certain users.
- Hiding fields on the form for use in a view or other computation.

NOTE

Hide-when is paragraph-level formatting. This means that if you use Hide-when on text on some part of a line, everything on that line will use Hide-when. Separate elements that should have different Hide-when settings should be placed in different paragraphs by pressing `Enter` **to move to the next line.**

Form Elements: Window Title

The Window Title appears in the Title bar of the document and in the Window menu when a user switches between Notes windows. If you do not set a window title, the word Untitled appears—which can be confusing for users. So it is usually a good idea to set a window title. You will use the Design pane of the form, as shown in Figure 2.19. While you are in Design mode on the form, click the background of the form. Then choose the Event drop-down menu at the top of the design pane and select Window Title. Type the formula or text to use for the window title using proper syntax. Note that the window title either can be static text or can change based on a formula. Enclose the text in quotation marks—e.g., "Survey Form"—to make the window title standard text.

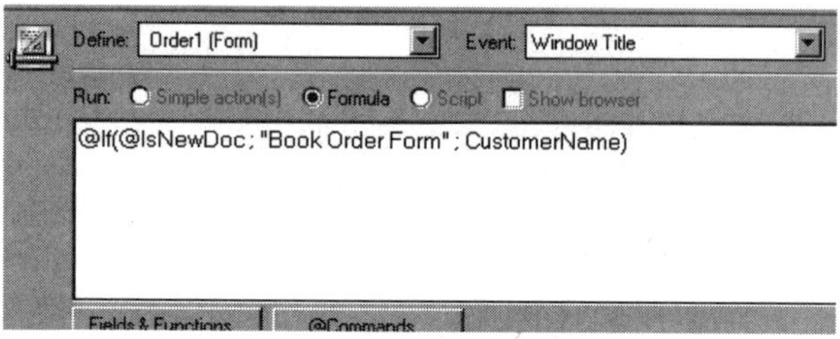

Figure 2.19 Create a window title.

Form Elements: Hotspots

A *hotspot* is a button, text spot, or graphic that performs an action. A variety of hotspots can perform actions such as linking to another document or URL, performing an action, running a formula, or creating pop-up text. Hotspots can be created on forms or on any document in a rich text field. To create a hotspot, select the text or graphic you want to make the hotspot (except for button hotspots), and choose Create . . . Hotspot. Then choose the type of hotspot you want to create. Hotspot types are described in the following paragraphs.

- *Link Hotspot.* A link hotspot will jump a user to a document, view, or database when the user clicks the hot text or graphic. Similar to a doclink, the link hotspot enables a more descriptive link than the doclink icon. To create a link hotspot, you must first select the document, view, or database you wish to link and choose Copy as Link. Once the link is on the clipboard, select the text or graphic that you want to make into the hotspot and choose Create . . . Hotspot . . . Link Hotspot. This feature is especially useful for cross-referencing other views, documents, or databases.
- *URL Link.* This option is new to R4.5.X and will not be tested on the exam. It enables the designer to type in any URL for the jump location of the hotspot.
- *Text Pop-up.* A text pop-up enables a small amount of text to be displayed in a pop-up dialog box when the user clicks on the text and holds down the mouse button. This command is only useful for a small amount of text, because there is no scrolling available in a pop-up. This feature is especially useful for definitions or footnotes. A text pop-up hotspot is shown in Figure 2.20.

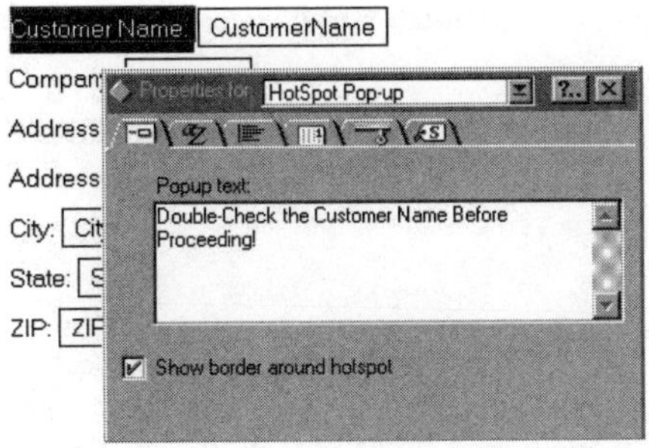

Figure 2.20 Text pop-up hotspot.

- *Button.* A hotspot button enables the user to perform an action, formula, or script from a button in a form or in any rich text field. To create a button hotspot, place your cursor where you want to place the button and either choose Create . . . Hotspot . . . Button from the menus or click the Create Hotspot Button SmartIcon. The Hotspot Button Properties InfoBox enables you to label the button and determine its width. The button label will determine its length. The width by default is two inches. If the button label will be long, you may also want to enable the option to wrap the text of the label. The button hotspot will then need to be given an action. This job will be done in the Design pane, using either an @Function, @Command, simple action, or LotusScript. Creating these actions will be discussed in Chapter 3. A button hotspot is shown in Figure 2.21.

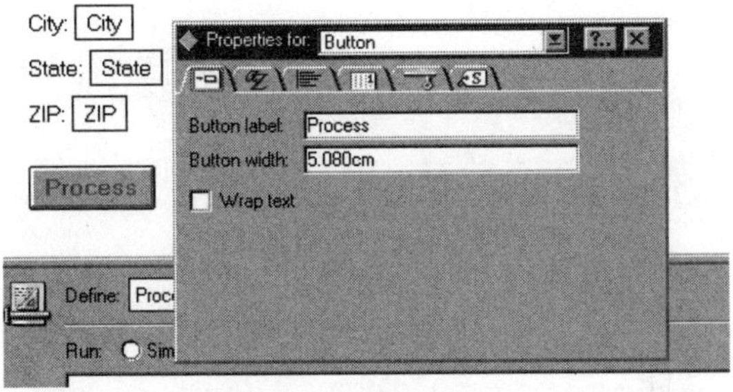

Figure 2.21 Button hotspot.

- *Formula Pop-up.* This command is similar to a text pop-up, but instead of typing in static text for the pop-up, the designer can create an action, formula, or script that resolves the text to be displayed.

- *Action Hotspot.* An action hotspot is similar to a button, but the action hotspot will be any text or graphic that can perform the indicated action, formula, or script.

EXAM TIP

For the Application Development I exam, be prepared for most of the questions on hotspots to concentrate on Link hotspots, text pop-up hotspots, or buttons. These types are defined as *manual hotspots.* The CLP exam guide, available from Lotus, lists manual hotspots on the list of competencies that you must know. Be prepared to know how to create these hotspots and define their actions.

Form Elements: Subforms

A subform is similar to any other form, in that it can contain all the same design elements. The purpose of a subform is to assist with consistency in forms by enabling multiple forms to share the same design elements. For example, if a company wants every form created in a database to share the company logo, a subform with this logo could be an easy way to insert it into all the forms. If a subform is changed, all forms using that subform will be updated to match automatically. To create a subform, choose Create . . . Design . . . Subforms. Options on the Subform Properties InfoBox are described in the next paragraphs. The Subform Properties InfoBox is displayed in Figure 2.22.

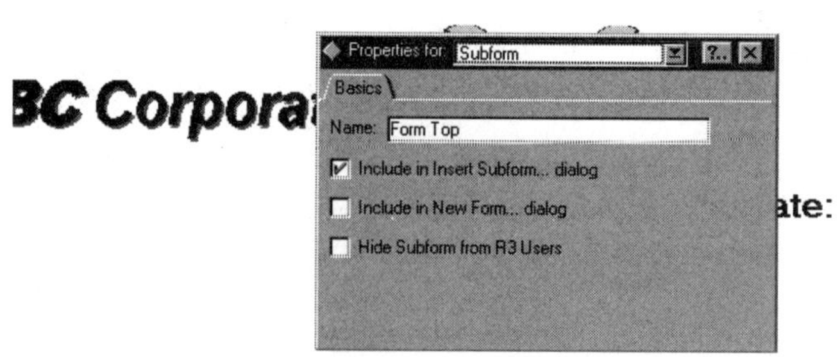

Figure 2.22 Subform Properties InfoBox.

Name. As always, it is important to give this design element a descriptive name.

Include in Insert Subform . . . dialog. This option, selected by default, enables the subform to be added to a form at design time. If this option is not selected, the subform can only be inserted through a function or formula into a form. The Insert Subform . . . dialog box is shown in Figure 2.23.

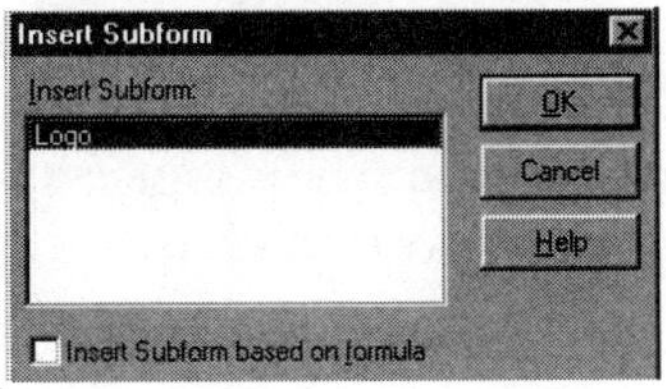

Figure 2.23 Insert Subform . . . dialog box.

Include in New Form . . . dialog. When this option is selected, the Insert Subform . . . dialog box is automatically displayed when creating a new form, with this subform listed as an option.

Hide Subform from R3 users. Some elements placed in subforms may not be compatible with R3. This option enables designers in a mixed environment to use the R4 function without displaying it to R3 users.

Sample Database

We will continue where we left off with our sample database in Chapter 1. The User Question Tracking database is now ready to have forms created inside it. The database needs two forms. The first form will enable the user to enter questions. The second form will be a response form used by the help desk personnel to answer the questions.

To create the main form, open the database and follow these steps:

1. In the User Question Tracking database, choose Create . . . Design . . . Form from the menus, or click the Create Form SmartIcon if you are in a design view.
2. Open the Form Properties InfoBox by choosing File . . . Document Properties from the menus or by clicking the Design Form Properties SmartIcon.

3. The form name should be Questions, with the synonym Q, as shown in Figure 2.24. Choose the Document form type, because this form will be the main form in the database. To ensure that the help desk is aware if the user changes the question after asking it, use Versioning. Choose the option Allow prior versions to become responses and choose to have versioning occur automatically at save time. It is important for the help desk to know who has questions, so do not allow the form to be anonymous. There are, however, two servers in the environment where the database will reside. Enable the databases to merge replication conflicts, therefore, on documents created with this form.

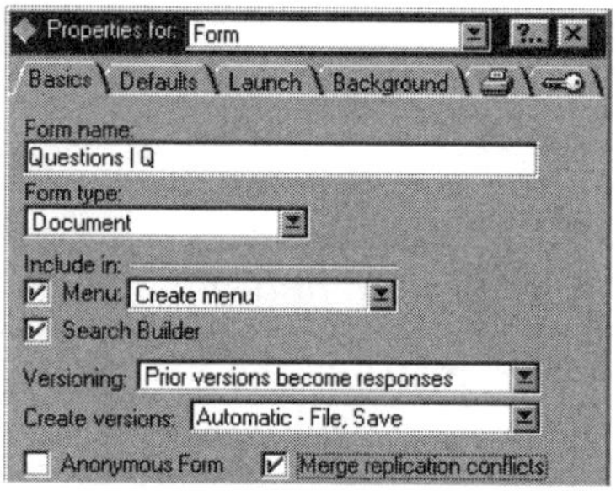

Figure 2.24 Sample database Basics tab.

4. This form should be the default database form, which you can choose on the Defaults tab, pictured in Figure 2.25. For now, leave all the other options on this tab at their default settings.

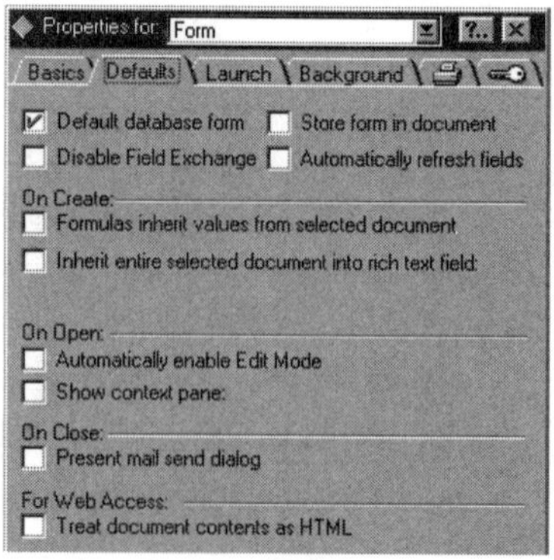

Figure 2.25 Sample database Defaults tab.

5. For the other tabs, there will be no changes from the defaults at this time. If you like, you can add color to the form background.
6. Save and close the form.

A second form will be in the database, as discussed earlier, for the help desk personnel to respond to questions. This second form will be a response-type form and will need to inherit some information from the main form. Create this form following the guidelines listed next.

NOTE
You might also want to create another form that would enable anyone to comment on the questions or answers. This form would be a response-to-response-type form.

1. Create the new form. Open the Form Properties InfoBox. The form should be called Answers, with the synonym A. Versioning will not be necessary in this form; however, this form should continue to have the capacity to merge replication conflicts. The Basics tab should look like Figure 2.26.

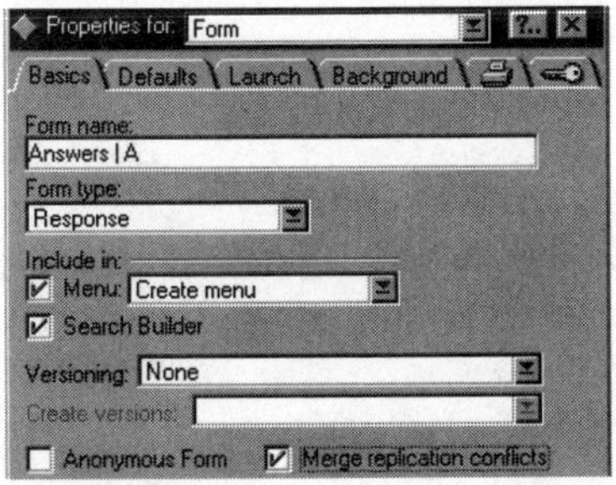

Figure 2.26 Response form Basics tab.

2. On the Defaults tab, you want to change options to accomplish the following tasks. First, you want this form to inherit some information from the main document to which it is a response. Second, to allow users to see their question with the answer, you want the parent of the document to be displayed for the user automatically. The Defaults tab should look like Figure 2.27.

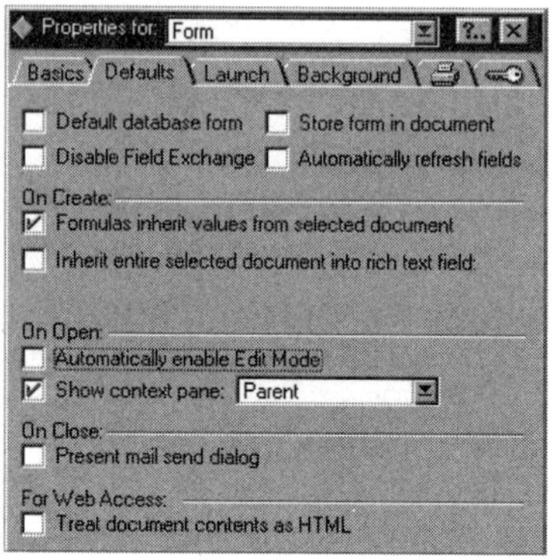

Figure 2.27 Response form Defaults tab.

3. Again, the other tabs will remain with their defaults, unless you would like to create a background color. Save and close the form.

Now that both forms exist, add static text and other design elements to the forms. The two following graphics show the forms as I have created them. Add the same design elements necessary to make your forms look similar.

Here are some hints on making the form by using the design elements discussed in the previous chapter. The logo and the date field are part of a subform called *Form Top*, because these elements will be reused on both forms. The sections were created as standard sections. Note there are two pieces of static text on the Answer form: *Followup* and *Notes*. These will be used by the support staff to track the status of the problem and should not be visible in the form when the person who asked the question reads it. These fields must be hidden from the reader using the basic Hide-when options. Figure 2.28 shows a Questions form, and Figure 2.29 shows the Answers form. Figure 2.30 shows the text properties for the followup and Notes static text set so that they will only be visible when editing.

Any other design work done on the form is done with static text and the Text Properties InfoBox. Remember that color, placement, and other formatting elements are not tested on the exam. Therefore, the form you see is simplistic—designed only to enable you to practice placing design elements on the form. Fields and formulas to enhance the form will be added in the next chapter.

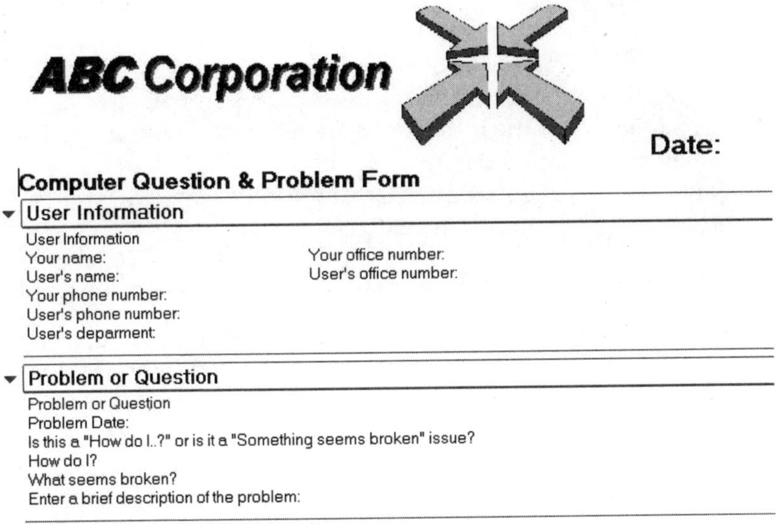

Figure 2.28 Questions form.

Figure 2.29 Answers form.

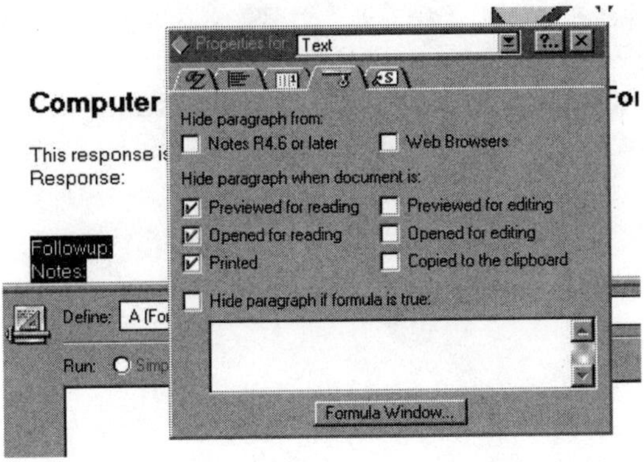

Figure 2.30 Answers form Hide-when options.

Review Questions

1. Guy is creating a database that will be translated into other languages. The translators do not know Notes and will not be able to change any of the formulas included in the application. They are only translating the text that end users will read. How can Guy ensure that any formulas using form names will still work after the application is translated?

   a. Choose the International Versioning option on the Form Properties InfoBox Basics tab.

   b. Allow the translators to give the form a new name, and then go back and edit the formulas with the new name.

   c. Use Find and Replace to change the name in formulas after the translators have given the form a new name.

   d. Use a synonym with the form name.

2. Milo created a database that uses seven forms. Only two of the forms, however, are used on a regular basis. The other five are used occasionally in special cases. Milo wants these forms to appear in a separate dialog box from the Create menu, not on the Create menu itself or on a submenu. How does he accomplish this?

   a. Set the Cascade under Other option on the Form Properties InfoBox.

   b. He cannot do this. He can only use the \ (backslash) to cascade form names, but they will still appear on the Create menu or on a submenu, not on a separate dialog box.

   c. Choose Include the Form on the Create . . . Other Dialog Menu in the Form Properties InfoBox.

   d. After two forms are created in a database, all additional forms are automatically placed in the Other . . . dialog box.

3. In the Inventory tracking database, the only time a user opens a document is when inventory needs to be checked out or returned. All other information is available in the database views. How can Chris make editing the documents easier for users?

    a. Automatically open the document in Edit mode using the option on the Form Properties InfoBox.

    b. Choose the Edit in View option on the Database Properties InfoBox.

    c. Enable the AutoLaunch options on the Form Properties InfoBox.

    d. Choose an AutoEdit form type on the Form Properties InfoBox.

4. Marcus is creating a database that contains the documentation for a piece of software his company is writing. Each time a particular document changes, Marcus wants to see the newest version as a main document in the view. He does not, however, want to lose previous versions of the documentation. What option can he select?

    a. New versions become responses

    b. New versions become siblings

    c. Prior versions become responses

    d. Save as Main document

5. Darren has created a sales database with six forms, which each relate to three different product lines. Darren wants to separate the forms by product line on the Create menu. How can he do this?

    a. Set the Cascade by Category option on the Form Properties InfoBox.

    b. Use a \ (backslash) in the Form name to separate the product line name from the name of the specific form. This command will create submenus based on the product line name.

    c. He cannot do this. He can only use the Create menu or the Create . . . Other dialog to list form names.

    d. Set the Create Form list option on the Database Properties InfoBox.

6. The form that Paul is creating has multiple long paragraphs of instructions for new employees. He knows, however, that many users do not need to see this. What feature can Paul use to allow users to decide whether they want to see this information?

   a. Hide-when options

   b. Standard Sections

   c. Controlled access sections

   d. Versioning

7. When Cortland sets up an evaluation database for professors at his college, students request that they be allowed to create evaluations without anyone (including a developer or system administrator) knowing which student gave which evaluation. How can Cortland accomplish this?

   a. Use the Anonymous option on the Database Properties InfoBox.

   b. Use the Anonymous option on the Form Properties InfoBox.

   c. Use Hide-When to hide the Author's field on the form.

   d. Leave off the Author's field on the form.

8. Melanie is creating a database for students at her college to ask questions, to which professors and administrators can easily and quickly respond. The professors want to see the question the student asked while they are writing their response. How can Melanie provide this?

   a. Show the Question document in a Parent context pane.

   b. Use versioning to let the professor answer the question on the same form.

   c. Use the Store Form in document option to display the Question form.

   d. Use inheritance to display the Question field on the Answer document.

9. Sean needs to be notified every time a user creates a new customer in the customer sales database. The first time a document is e-mailed to him from the customer sales database, he opens it in his e-mail file but cannot read any of the data. What option did the designer forget to enable so that Sean could read the documents correctly?

   a. Send Form with Mail
   b. Enable Field Exchange
   c. Store Form in Document
   d. Allow use of Stored Forms

10. Cindy wants to create five forms in her database that all have the same look and feel, including company logo, author's name, date, and subject. She also knows, however, that some of these elements may change stylistically. What Notes feature will she use to do the least amount of work in creating these forms?

    a. She will copy and paste the logo and other design elements from one form to another.
    b. She will place the logo, text and fields on a subform that can be inserted into each form at create time.
    c. She will enable inheritance, so that when she creates each form, it inherits information from the other forms.
    d. She cannot do this. Each form has to be created separately with all unique design elements.

11. When creating a form, Aaron wants to allow users to link to other documents in the database that explain the questions on the form. He does not want to use the default doclink icon, however. What is an easy way for him to allow users to link to the other documents?

    a. Create a Link Hotspot
    b. Create a Button Hotspot
    c. Create a doclink, then choose Create . . . Other . . . Use Other Icon.
    d. In the Object Properties InfoBox for the doclink, choose the Icon drop-down list to display a different icon to the user.

12. Stephanie wants to create a form in her company suggestion database that can be used by the executive committee to respond either to the suggestions (the default form in the database) or to the comments (used to respond to the suggestions). She calls the form *Executive Comment*. What type of form must she use?

    a. Descendant form

    b. Response form

    c. Child form

    d. Response-to-response form

13. John and Aaron both edit all the documents in the *Surf-n-Sip* newsletter. They are using different replicas of the database: one in Houston and one in Germany. What can Joe do to help eliminate some of the replication conflicts they are complaining about?

    a. Put John and Aaron on the same server. Enable the document lock option on each of the forms in the newsletter database, so that only one of them can edit the documents at a time.

    b. Continue to use both databases. In Database Properties, enable the Merge replication conflicts option.

    c. Place both John and Aaron on the same server. In Database Properties, enable the Merge replication conflict option.

    d. Continue to use both databases. In Form Properties, enable the Merge replication conflict option.

14. To make the forms in her recipes database easier to use, K.C. wants to organize the ingredients, preparation, and cooking portions of the form so her users can selectively hide the parts they do not need. What function of Notes should she enable?

    a. Hide-when

    b. Standard Sections

    c. Controlled access sections

    d. Subform

15. Julie created a discussion database to help her users decide issues and challenges in their organization. To help users respond to the issues, she wants to have the response document display the original question. What does she need to do to accomplish this?

   a. Enable Inheritance on the Database Properties InfoBox and use fields with the same names on both forms.

   b. Enable Inheritance on the Form Properties InfoBox and use fields with the same names on both forms.

   c. Use only a response-to-response form in the database and use Hide-when to determine what sections are displayed.

   d. Enable Inheritance on the Form Properties InfoBox and use fields with the default value of the field name from the main form.

16. Your new developer has two fields, *name* and *section,* with two static texts describing each field on the line above the fields. The developer tried to use Hide-when options to hide *name* for editing while *section* will always be visible. When he shows you the form in Read mode, however, both of the fields and the static texts still show (see Figure 2.31).

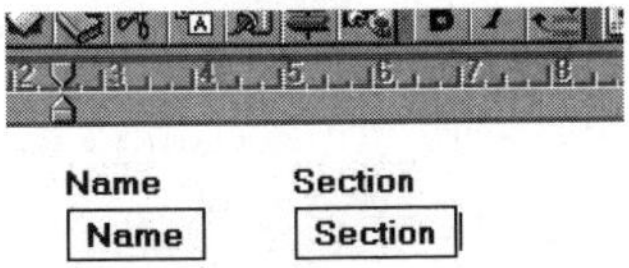

Figure 2.31 Name and Section.

   What can you change to help the developer's form work correctly?

   a. Place the static text on different lines and redefine the Hide-when settings.

   b. Place the field on the same line as the static text and redefine the Hide-when settings.

   c. Place the static text on two different subforms and redefine the Hide-when settings.

   d. Place each static text/field combination on different lines and redefine the Hide-when settings.

Answers

1. d. Use a synonym with the form name.
2. c. Use the Create . . . Other option in the Form Properties InfoBox.
3. c. Enable the AutoLaunch options on the Form Properties InfoBox.
4. c. Prior versions become responses.
5. b. Use a backslash in the form name.
6. a. Hide-when options
7. b. Use the Anonymous option in the Form Properties InfoBox.
8. a. Show the Question document in a parent context pane.
9. c. Store form in document
10. b. Create the elements in a subform that can be inserted into each form.
11. a. Create a link hotspot
12. d. Response-to-response form
13. d. Continue to use both databases. In the Form Properties InfoBox, enable the Merge replication conflict option.
14. b. Standard sections
15. d. Enable Inheritance on the Form Properties InfoBox and use fields with the default value of the field name from the main form.
16. d. Place each static text/field combination on different lines and redefine the Hide-when settings.

CHAPTER 3

Fields, Formulas, Agents, and Actions

After you have created a form and placed the necessary static text and other design elements on the form, you are ready to give the form some action by adding fields, formulas, agents, and other actions. Without these elements, your form is nothing more than text and graphics; there is nothing for a user to do. A field is the basic, smallest element on a form. It contains the data, either entered by a user or determined by the program. A variety of field types are associated with a type of data, ranging from text to numbers to dates and times. The exam covers both the ability to add fields of all varieties to a form and the ability to add formulas to fields for computing data. These will be covered in this chapter in three steps. First, we will define the characteristics of a field using the Field Properties InfoBox. Second, we will cover formulas, including @Functions and @Commands and their appropriate syntax. Finally, we will discuss the ability to use agents and actions to provide additional functionality.

Chapter Objectives

You should be able to answer questions based on the following objectives after reading this chapter:

- Add fields to a form.
- Provide keyword choices in a field.
- Use required fields on forms.
- Apply proper case or other formatting to text a user has entered in a field.
- Prepare a form to use categories.
- Use shared fields.
- Use fields that contain multi-values.
- Use @Functions and @Commands in formulas.
- Write/Use manual agents.
- Write/Use Actions in forms.

Field Properties InfoBox

The major defining characteristics of a field are its name, its data type, whether it is computed or editable, how it is displayed to users, and whether it contains formulas. These characteristics are defined by the Field Properties InfoBox. The following paragraphs describe the options available in the Field Properties InfoBox, shown in Figure 3.1.

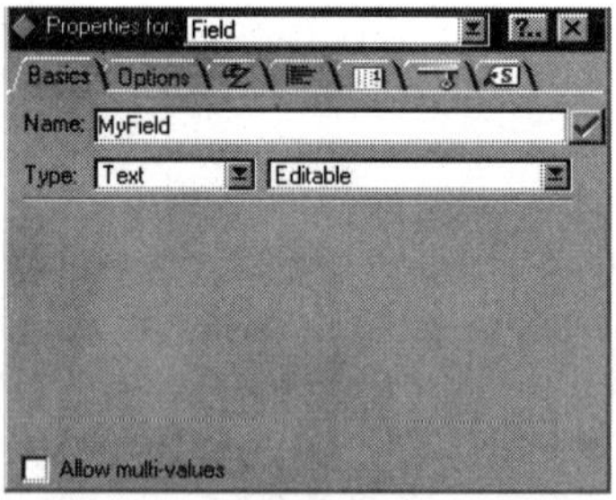

Figure 3.1 Field Properties InfoBox.

To insert a field onto a form, place your cursor in the form where you would like to insert the field. To insert the field, choose either the Create Field SmartIcon, or Create . . . Field from the menus. Either choice inserts the field and displays the Field Properties InfoBox automatically. As discussed in the last chapter, it is useful to build the static text and other design elements of the form first to give the form some shape and to remind yourself of the organization of the form. In addition, it is a good idea to draw out all your forms, including static text, design elements, and fields, before beginning to create your form in Notes. This prevents you from forgetting the name or function of fields as you add more of them to your form.

The Basics Tab

The Basics tab on the Field Properties InfoBox enables you to set the defining information about the field, such as field name and field type. The Basics tab changes dynamically based on the choices you make for field type. The fields on the Basics tab are described in the following paragraphs:

- *Name*. Each field should be given a descriptive, unique name. No two fields can have the same name on the same form. The length of the field name is limited to 32 characters. When naming a field, no spaces are allowed (use dashes, underscoring, and capitalization instead). In addition, field names cannot begin with numbers and should not begin with dollar signs ($) or at symbols (@), as these symbols have specific, reserved meanings in Notes. Finally, certain field names are reserved for use by Notes, such as SendTo, CopyTo, Sign, and Form. These names should never be used for purposes other than those defined by Notes.
- *Type*. The following choices are available for field type: text, time (which includes both time and date), number, keyword, rich text, authors, names, and readers. In addition, for each type of field, you will have to make the choice as to whether it is editable, computed, computed for display, or computed when composed. These choices are described further later in this chapter. In addition, remember that the choices you make in the type field will change the rest of the options available in the dialog box.
- *Allow Multi-values*. This option, which appears with most of the field types, allows multiple values in a particular field. The user will be able to create lists in the field, for example. In addition, multi-value options on the Options tab will be available for editing. This field allows multiple values, but Notes restricts each field to a single type of value. If a field is a text type, numeric values will have to be converted to text to be allowed in the field. Only one type of value is allowed in a field.

FIELD TYPES

The field types and the options that relate to them are discussed here. For the exam, it is important to know each of the field types and their use, although the exam concentrates on text, rich text, and keyword field types:

- *Text.* This field type has the fewest options and is very common for simple user input. In this field type, the user can type letters, punctuation, spaces, and numbers (unless they are used mathematically or derived from a formula). The user cannot change the formatting of this field type; it is set by the developer.
- *Time.* The time field includes both time and date data. The additional fields displayed by this field choice are shown in Figure 3.2. These include Show, Date Format, Time Format, and Time Zone. The show field allows you to determine whether you want to display time and date, date only, time only, or Today and time. The option for Today and time creates a formula pulling in the current date from the operating system. The Date and Time formats give you a variety of options for displaying the date and time. Internally, Notes stores time and date in the format `MM/DD/YYYY HH:MM:SS`. This provides complete Year-2000 compliance, as all four-year characters are stored internally and used in calculations.

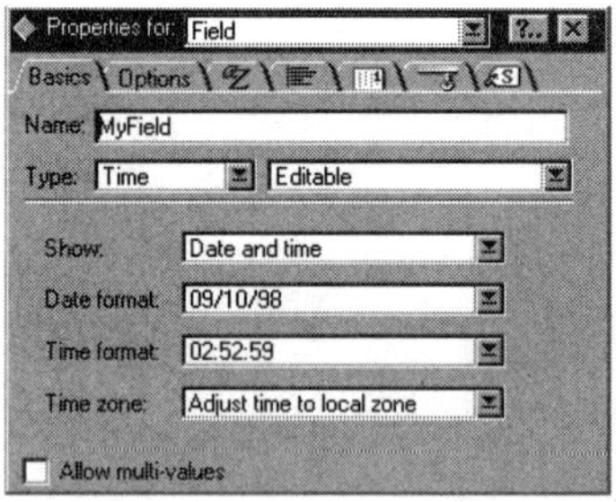

Figure 3.2 Basics tab, Time field type.

- *Numbers.* The numbers field type is used for numeric values that will be used in calculations. Numbers can be displayed as general numbers, currency, scientific, or fixed. In addition, there are options for percentages, punctuation, and negative numbers, as displayed in Figure 3.3.

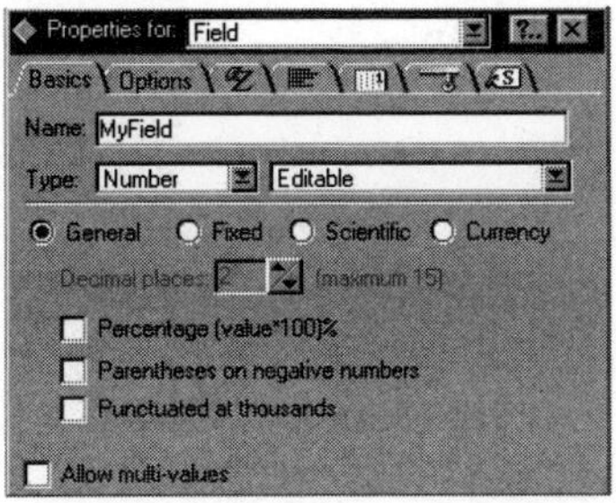

Figure 3.3 Basics Tab, Number field type.

- *Rich Text field type.* Rich text is similar to a text field but gives a user the capability to insert formatted text, doclinks, attachments, hotspots, graphics, and other objects. There should normally be only one rich text field per form. Note that rich text field data generally cannot be displayed in a view or used in a formula.
- *Authors field type.* The Authors field determines who can edit a document after it has been saved. Adding John's name to the Author's field, for example, gives him the capability to edit a document he may or may not have created. John must have author access to the database for this to work. This relates to database security and will be discussed in more depth in Chapter 11.
- *Readers field type.* The readers field restricts read access to a document, regardless of the user's level of access in the database ACL. This field relates to database security and will be discussed in more depth in Chapter 11.
- *Names field type.* The names field type will display user or server names as they would appear in a Notes ID file. It displays a hierarchical name in a distinguished name format rather than the canonical format stored internally. Instead of displaying `CN=Libby Schwarz/OU=Authors/O=TotalSem`, for example, this field would display `Libby Schwarz/Authors/TotalSem`. This type of field is not used for security.
- *Keywords.* The keyword field type provides the user with a list of choices to complete the field. This allows the designer to enforce consistency and limit the responses to a question to the types of answer he or she wants. The keyword list can be generated either by typing in the list or programmatically. These options are discussed below. Note that the keyword field type is the most heavily

tested type on the exam. In Figure 3.4, the field type is keywords, which changes not only the options on the Basics tab but also adds a tab to the dialog box.

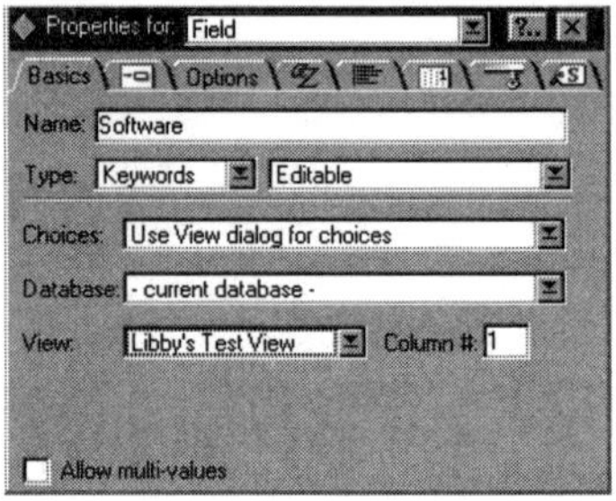

Figure 3.4 Basics Tab, Keyword field type.

- *Enter choices (one per line).* This allows the developer to enter the choices manually. These static text choices will always be the same when presented to the user. The design would have to be changed to add or delete a choice. To allow the user to add a choice other than the listed keywords, however, select the option to Allow values not in list. This gives the user the option to add an item to the keyword list. If the list of choices will change very little or remain static, this is a good choice. If the list will be changing often, however, it may be easier for the designer to use one of the other options.

NOTE
Any items that are added to the keyword list using this option are not added to the list permanently. A formula must be used to add items to the list using user input.

To enter keyword choices, type the choices in the window, pressing Enter after each choice. You have the option of using synonyms in the keyword lists, similarly to the synonyms discussed for form names. Figure 3.5, shows a keyword list that uses synonyms. To use a synonym in the keyword list, type in the keyword option that you want displayed to the user, type a pipe character (|), and then type the synonym that you want to use internally. In keyword lists, the most common synonym is numeric. One advantage of using keywords is the ability to change the name of the option presented to the user without having to change any formulas that use the keyword.

NOTE

When keywords are used in column formulas in Views, the synonym is displayed by default rather than the keyword normally displayed to the user. In our example, the view would display the numbers zero through two rather than yes, no, or maybe. Later in the book we will discuss ways to display the keyword rather than the synonym.

Figure 3.5 Keyword list with synonyms.

- *Use Formula for choices*. When you use a formula to determine keyword choices, you can allow the choices to change dynamically. For example, assume your keyword list is for placing an order. You would only want the items that are actually available to be shown in the keyword list. This formula might be an @DbColumn formula that allows the values to be generated from a view. Alternatively, the list of values might come from outside the Notes application. This is not tested on the Application Development I exam.
- *Use Address Dialog for choices*. Using the Address dialog presents the user with a Names dialog box that allows the user to choose from an address book (either personal or public). To speed up lookups, you can choose to enable the *Look up names as each character is entered* option. This emulates the lookup in mail memo documents.
- *Use Access Control List for choices*. This presents the user with a dialog box that pulls groups, users, servers, and roles from the Access Control List for the database. Use this option when the users with access to the database provide the best choices for the field.
- *Use View Dialog for choices*. This option is new to R4. It allows the designer to specify a view, including database, view, and column, from which the user can make choices. Figure 3.6 shows what the user sees when he clicks the arrow for choices in this type of list.

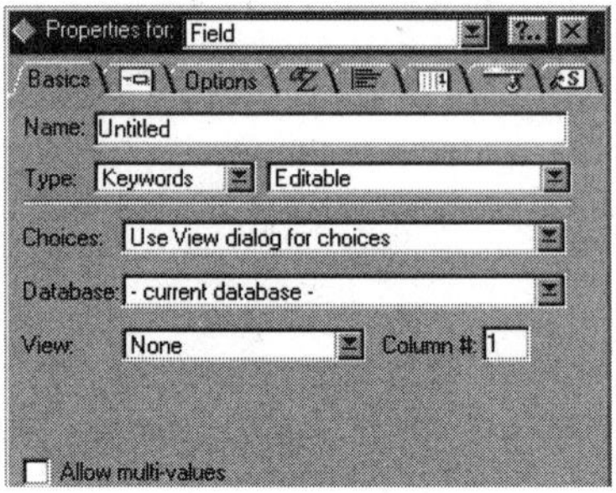

Figure 3.6 Keyword from View dialog.

Choosing the Keyword data type also requires deciding how the keyword list should be displayed to the user. This choice is made on the Keyword Option tab, shown in Figure 3.7. This tab is only available for a Keyword data type. The choices available on this tab are described in these paragraphs.

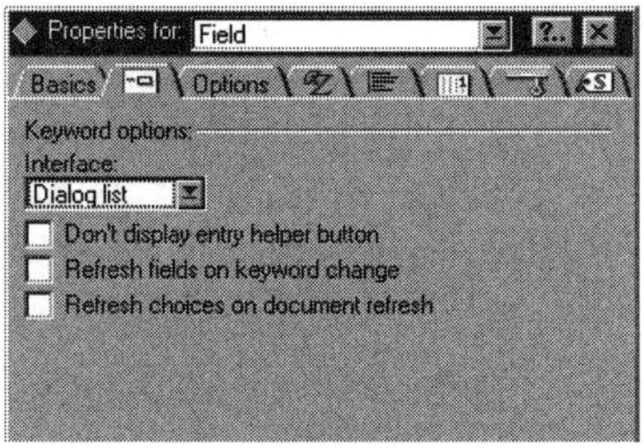

Figure 3.7 Keyword Options tab.

- *Interface.* The interface option enables the designer to determine whether the user will see a dialog list, checkbox, or radio buttons. If the designer chooses a dialog list, the interface the user sees will look like Figure 3.8. If *Allow Values Not in List* is enabled, the user can type in new keywords at the bottom of the dialog list. When the designer chooses checkbox as the interface type, the interface will look like Figure 3.9. Please note that multi-values are automatically assumed when using checkboxes but allow values not in list is not allowed. Checkboxes have frame formatting and number of columns available as options. Finally, if the designer chooses radio buttons, as shown in Figure 3.10, the user has the option to pick only one choice from the list. With this option, neither multi-values nor values not in the list are allowed. The same frame and column options are available for checkboxes.

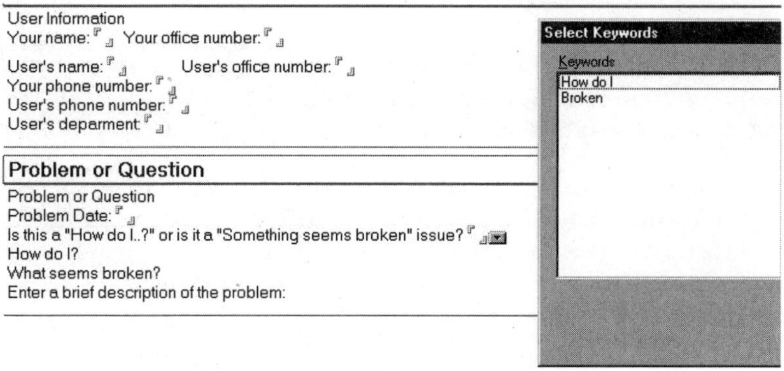

Figure 3.8 Dialog list display.

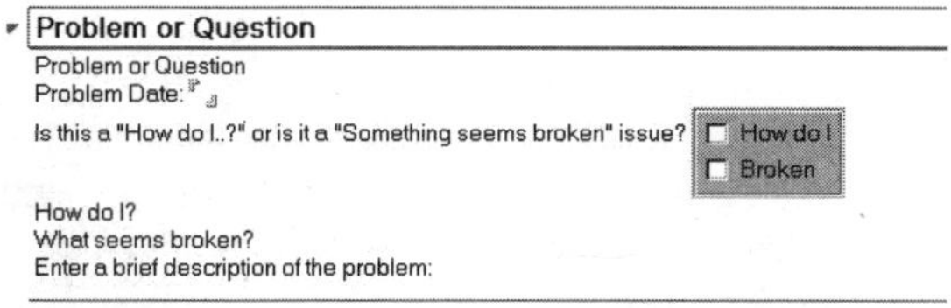

Figure 3.9 Checkbox display.

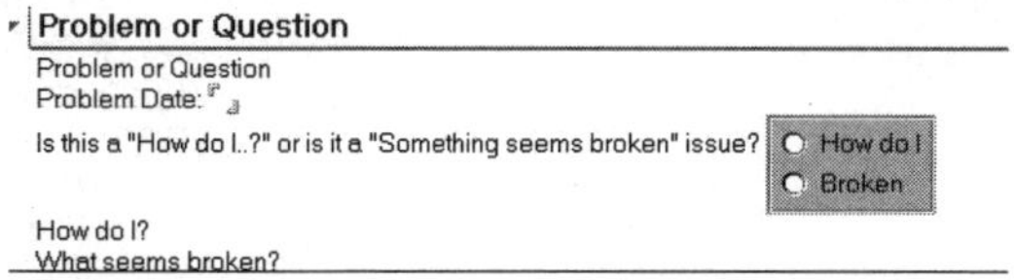

Figure 3.10 Radio Button display.

- *Don't display entry helper button.* This choice, available for dialog lists, removes the list button that is automatically displayed next to the field. This forces the user to type the correct option or press Ctrl+Enter to see the list.
- *Refresh fields on keyword change.* Use this option if other formulas in the form are dependent on the field's value. For example, if you want to hide paragraphs depending on the value of a keyword field, you need to have this option enabled.
- *Refresh choices on document refresh.* This option refreshes the keyword list each time the document is refreshed. This is necessary if the keyword list is created from something other than a statically typed list, as the keywords may change.

Editable versus Computed Fields

Most field types, with the exception of Rich Text, have four additional options for how data will be entered into the field. These options are described in the following paragraphs:

- *Editable.* The user manually enters information into an editable field. There may be a default value entered into the field by formula when the document is created. To accept the default value, the user simply does not type an entry into the field.
- *Computed.* When documents are created, refreshed, or saved, the calculated value of a computed field is recalculated. The field is completed by a formula entered by the designer. The end user cannot change a computed field.
- *Computed for display.* In a Computed for display field, the value is recalculated every time the document is opened for reading or editing or any time the document is refreshed. The value of the field is not stored when the document is saved and closed. Because this data is not stored when the document is closed, the data in this field cannot be displayed in a view. This type of data will only be displayed for viewing in an open document.
- *Computed when composed.* A computed when composed field is only calculated once, at the time the document is created. It is never recalculated. This field type is useful for a date stamp on a document, for example, or a field showing the original creator of the document.

The Options Tab

The Options tab has choices related to field-level help and multivalue options, as shown in Figure 3.11. These choices are described in the following paragraphs:

- *Help description.* You can type in a brief description of how the user should use the field by using this option. The help will be displayed in the Status bar at the bottom of the workspace, if the user has turned on field help, using View . . . Show . . . Field Help.
- *Give this field default focus.* The field that has this option enabled will be where the user's cursor will go automatically when the document is opened in edit mode.

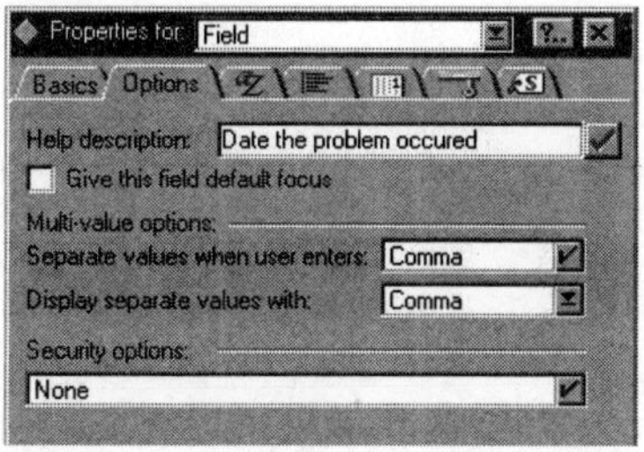

Figure 3.11 Options tab.

- *Multi-value options*. These options are grayed out unless the Allow multi-values option is enabled. The two choices in this section are for Separate values when user enters and Display separate values with. The choices for both options are the same and include the following: space, comma, semicolon, new line, and blank line. Note that when the field is a keyword field, these options are not available.
- *Security options*. This field allows the designer to secure the field with signing, encryption, and ACL. However, the security options are not tested on Application Development I and will be discussed in Chapter 11.

The Font Tab

The Font tab, shown in Figure 3.12, is similar to the Font tab shown in other InfoBoxes. In general, this tab allows the designer to set the font for anything that is affected by font, including static text, section titles, and fields. The Field properties Font tab sets the default font for the field display. This defines the font users type into fields unless it is a rich text field. A rich text field retains the default font, but the user has the ability to change the font information.

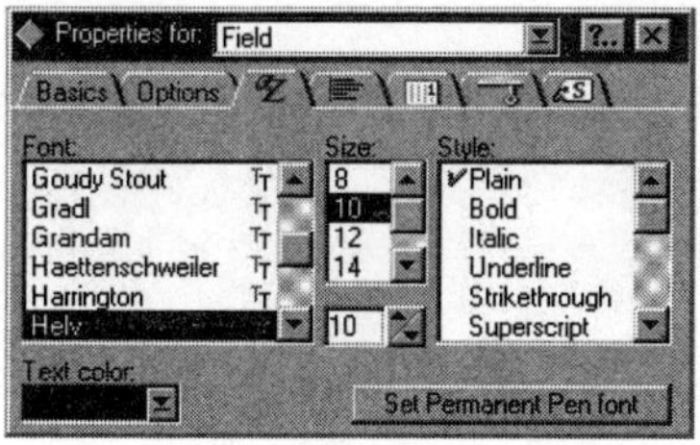

Figure 3.12 Font tab.

The Spacing Tab

The Spacing tab, shown in Figure 3.13, is again similar to the Spacing tab shown in other InfoBoxes. This tab allows the designer to set spacing, alignment, and list defaults for the paragraph. This tab is not tested on the exam.

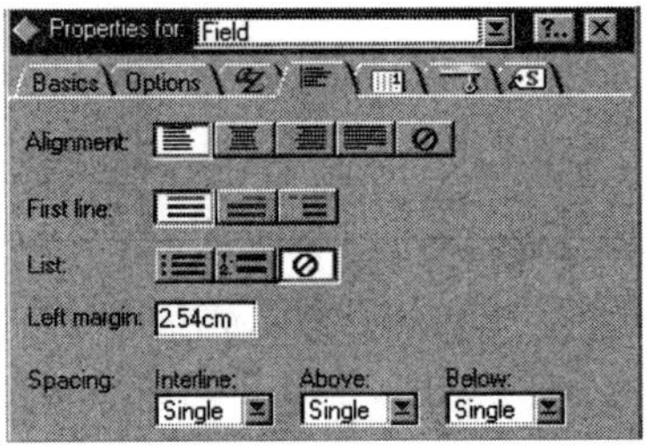

Figure 3.13 Spacing tab.

The Pagination Tab

This tab, shown in Figure 3.14, enables the designer to define the tabbing and pagination of the form. This tab is not explicitly tested on the exam.

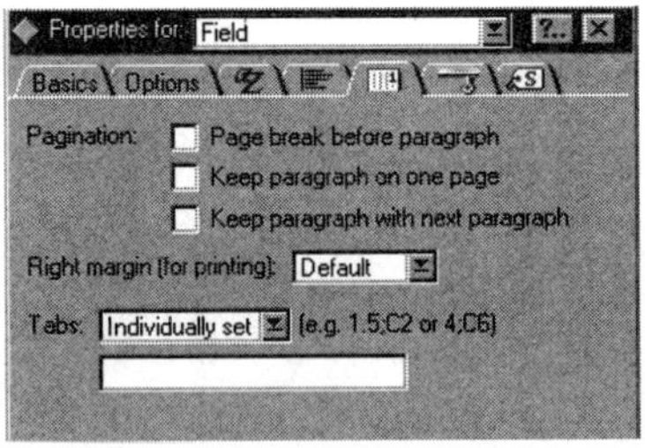

Figure 3.14 Pagination tab.

The Hide-When tab has been described earlier, in Chapter 2, and as you can see from Figure 3.15, is the same in this InfoBox.

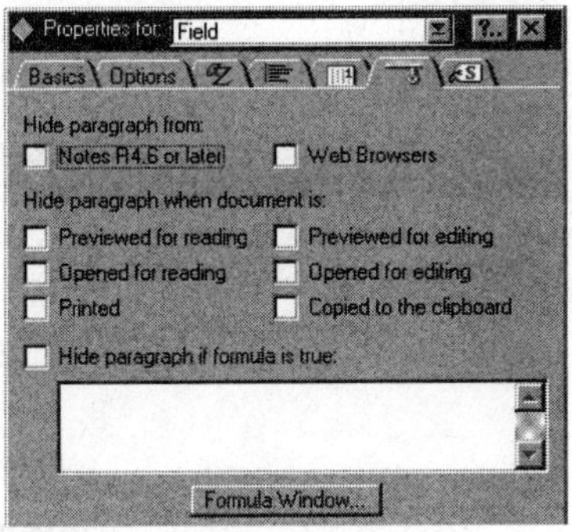

Figure 3.15 Hide-When tab.

The Style Tab

Like the Font and Pagination tabs, the Style tab relates to the formatting of the text and paragraphs on the form and is not directly tested on the Application Development I exam.

Additional Field Types

Notes contains two additional field types that are not listed in the Field Properties InfoBox. These two field types are Shared fields and Categories fields. These field types enable additional functionality in the Notes database.

Shared Fields

Shared fields, like subforms, enable a Notes design element to be reused on multiple forms within a database. The main difference between the two is that a subform enables a group of design elements, including text, graphics, and fields, to be reused. A shared field allows a particular field to be reused from form to form. Shared fields can be created using the following methods:

- Create a new shared field by choosing Create . . . Design . . . Shared Field from the menus. This displays the Shared Field Properties InfoBox. The options on this InfoBox are the same as those for any other field.
- Share an existing single-use field by choosing Design . . . Share this field. This automatically shares the field and makes it available in the Insert Shared Field dialog box. There is no option to

stop sharing a field; instead, to stop sharing, cut the field and then paste it back onto the form. It should be a single use field when you paste it.

- Copy shared fields from another database by using the Design . . . Shared Fields view. Copy the shared fields you want to use in the new database and then paste them into the Shared Fields view in the new database.

To insert an already created shared field onto a form, choose Create . . . Insert Shared Field. This brings up the Shared Fields dialog box, shown in Figure 3.16. Choose the shared field you want to insert and click OK. It is important to remember that any change made to a shared field affects that field on every form in the database. Using shared fields, like using Subforms, saves development time and helps to ensure consistency between forms in the database. This makes the database easier to use.

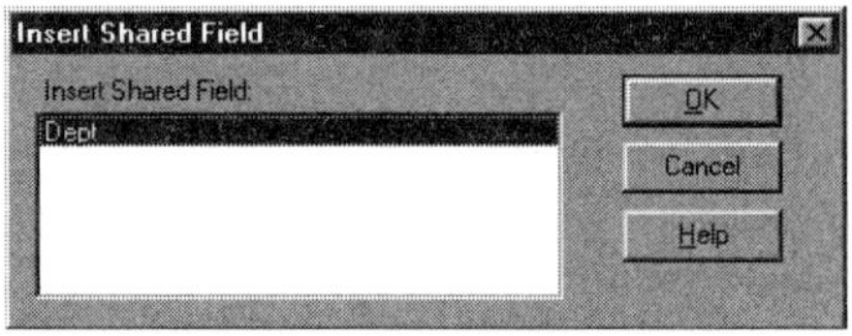

Figure 3.16 Shared Fields dialog box.

Categories Fields

In many Notes views, the documents are organized by Category, as shown in Figure 3.17. To use the categorizing functionality of Notes, a Categories field must exist on the form you want to categorize. A Categories field enables the user to use the Action . . . Categorize menu option if the designer has created a View that uses the categories.

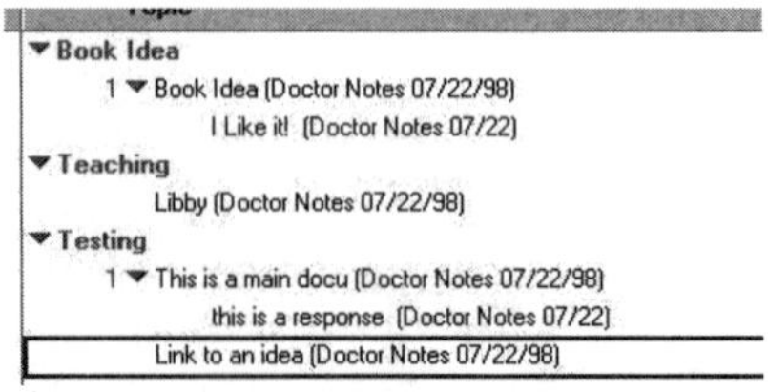

Figure 3.17 Categorized View.

Often the Categories field will be a keyword field that enables the designer to specify a list of categories. The value of the Categories field then will be used to group documents that have the same purpose or content, based on the selected category.

Formulas in Fields

In all fields, whether editable or computed, the designer has the option to define the value using a formula. The designer creates a formula with one of the available programming languages. There are two main types of programming languages in Notes: the formula language discussed in this chapter and LotusScript. LotusScript is an object-oriented programming language that is similar to C++ or Visual Basic. LotusScript is not covered in the Application Development I and II exams and is, therefore, not covered in this book.

The formula language uses elements such as constants, variables, keywords, and @Functions. Formulas are one of the basic elements of creating applications, appearing in everything from Hide-When elements to columns to fields. This chapter describes how to create and use formulas, as they are covered on the exam, including syntax. You must be able to evaluate a formula in the same way Notes will for the Application Development I exam. However, this section is not intended to be a complete reference on using formulas or any of their elements. For more information about formulas, especially if you have never written formulas in Notes before, use the Lotus Notes Help documentation to supplement this guide.

Formula Components

The first thing to know about a formula is how to put it together. To put formulas together correctly, you have to use the correct syntax and the correct formula elements. The correct syntax includes elements such as separators, case, parentheses, and quotation marks. These syntax elements are described in the following paragraphs. The syntax elements are especially important for troubleshooting questions on the Application Development I exam. The different elements of a formula that will be joined together using this correct syntax are called components. The following are the components of formulas in Notes: constants, variables, operators, keywords, @Functions, and @Commands. Each of these components is discussed in the following sections.

Syntax

Syntax is one of the most important elements of application development and of passing the Application Development I exam. The exam expects that the developer can correctly apply the following general syntax rules.

- *Use separators correctly*. The colon (:) is used to separate list items. A semicolon (;) is used to separate multiple statements or arguments. Both colons and semicolons are also used to end lines. Spaces are ignored in most cases with the following exceptions. First, all keyword formula components must be followed by at least one space. In addition, in text constants, include spaces where they are appropriate.
- *Use case correctly*. Most elements of Notes are not case-sensitive. Keyword formula components, however, should be used in uppercase. In addition, most field names, @Functions, and @Commands are in used in mixed case.
- *Use parentheses when necessary to control the order of evaluation and to enclose @Function arguments*. Remember to use the same number of open and close parentheses in complex formulas.
- *Use quotation marks to enclose text strings in formulas*. If you enclose a field name in quotation marks, it will be treated as a text string. To use quotes in a text string, place a \ before them, as in "Libby\'s".
- *Use a single data type in all formulas*. To combine data types in formulas, use conversion @Functions, such as @Text and @TextToNumber.
- *When using logical operators, ensure that each expression can be evaluated individually*. Each expression must result in a True or False result without elements from the other side of the expression. The expression Select form = "Question" | "Answer" is incorrect, for example. Use the expression Select form = "Question" | form = "Answer" instead.
- *To use time values as text, enclose them in quotation marks; to use time values as time constants, enclose them in square brackets.*

Constants

Constants can be any piece of static text including numbers, times, and words. These elements cannot change. They must be surrounded by quotes to use them in formulas, such as "Constant."

For a mathematical comparison, in the equation 2x-y, the 2 is a constant.

Variables

Variables in Notes are most often field names, but they can also be temporary variables. Field names used in formulas can be used alone or as part of an @Function formula to create a new result. When using field names as variables, most often they will be used without quotes. The option is usually available to pick the field name from a list while creating a formula. This is highly recommended, as most of the mistakes made in formulas are syntax and typographical errors. Temporary variables are set only inside a formula and are used to store information only while the formula is being evaluated. In our mathematical comparison of 2x-y, both the x and y are variables.

Operators

Operators perform the actions on the constants and variables in a formula. In our mathematical comparison, the - (minus sign) is the operator. The operators available in Notes are listed below in the order in which they are evaluated by Notes.

Table 3.1 Notes Operators

Assignment	:= (Colon, equal)	UserName := Libby (assigns value of Libby to the field UserName)
Concatenation	: (colon, used for lists)	"Apples" : "Cherries" : "Strawberries"
	+ (plus, used for text)	UserName + "developed this database." returns: Libby developed this database.

continues

Table 3.1 Continued

Arithmetic	* (multiplication)	TotalPrice*8.25
	/ (division)	NumberOfBooks/Price
	+ (addition)	TotalPrice + Tax + Shipping
	- (subtraction)	NumberOfBooks - DiscountedBooks
Comparison	= (equal to)	SELECT Form = "Questions"
	< (less than)	SELECT NumberOfBooks < 50
	> (greater than)	SELECT NumberOfBooks > 100
	!=, =!, <>, >< (not equal to)	SELECT UserName != "Libby"
Logical	& (And) | (Or) ! (Not)	SELECT Form = "Questions" | Form = "Answers"

NOTE
Concatenation is used with text strings; addition is used with numbers.

NOTE
The operators | (or) and & (and) sometimes confuse new developers. Using & (and) between logical expressions means that both expressions must be true for the whole expression to be true. Using | (or) between two expressions means that the whole expression is true if *either* part is true. Or is less limiting; it will return more true values. And is more limiting; it will be true less often.

NOTE
The equal sign (=) is used in Notes for comparison. MyName is equal to (=) UserName. It does not set something equal to something else—that is an assignment function carried out by the symbol := (colon plus equal sign).

This order of evaluation is subject to the normal mathematical order of evaluation: parentheses, exponents, multiplication, division, addition, and subtraction. For example, in the mathematical formula 2+(5-3), you would evaluate the 5-3 first and then add the result to the 2. Formulas in Notes will be evaluated the same way, such as @UpperCase (@Left ("Libby and Brian"; 5)). Notes evaluates the @Left function first and then evaluates the @UpperCase function.

Keywords

The formula Keywords described below perform special functions within Notes formulas. You will notice that they are capitalized here. The capitalization is a convention for ease of reading the formulas, not a requirement. If you do not capitalize them in formulas, however, Notes capitalizes them automatically.

NOTE
Do not confuse formula keyword components with keyword field types. They are separate, unrelated formula elements.

- *SELECT*: This keyword is used to choose a value or form for an action. Most often, you will see the SELECT keyword used in selection formulas. If a SELECT keyword is needed in a formula but has been left out by the designer, SELECT @ALL will be appended to the formula automatically. For example, SELECT Form = "Answer" in a formula would choose all documents created with the form Answer. SELECT is also used often to evaluate keyword values for Hide-When formulas.
- *DEFAULT*: The DEFAULT keyword is used to set a value in a field. If a field does not exist in a document, Notes will evaluate the document as though it does exist, with the specified value. Take, for example, DEFAULT MyName := Author. This formula sets the field MyName equal to the Author of the document.
- *FIELD*: This keyword is similar to DEFAULT, in that it sets a field value. If the field does exist, its value is replaced with the indicated value; if the field does not exist, it is created with the assigned value. For example, in FIELD CarColor := "Blue," the value Blue is assigned to the field CarColor.
- *REM*: As in many programming languages, the REM statement allows the designer to make *remarks* about the formula. Anything immediately after the REM keyword but on the same line, will be

ignored and not evaluated as part of the formula. It is usually used to document the purpose of a formula or program.

- *ENVIRONMENT*: This not tested on the Application Development I exam. Its basic purpose is to allow the developer to assign an environment variable in either the NOTES.INI file (Windows) or the Preferences file (Macintosh).

@Functions

An @Function (pronounced *at function*) is a predefined Notes formula used to perform specified actions. They allow the developers to add functionality to the applications without a complete object-oriented programming language (although this is provided in LotusScript). There are approximately 150 to 200 @Functions, approximately 50 to 75 of which may be tested on the Application Development I exam. The exam focuses on some of the @Functions that are new to release 4, as well as some of the commonly used building blocks. This section describes some of the @Functions tested on the exam including their use, syntax, and examples.

Here you will find a short list of examples to help you understand some of the syntax rules and uses of the @Functions you are expected to know for the Application Development I exams, often known as Level 1 @Functions. This is not a comprehensive or exhaustive list—for more information on @Functions, there is a more complete list in Appendix A to help you to study. Additionally, the Notes User Help database has a complete listing, including examples, of all available @Functions. In addition, when choosing an @Function from the lists available in all Notes formula windows, press F1 while any @Function is selected for help on that @Function. Figure 3.18 displays the @Function list dialog box.

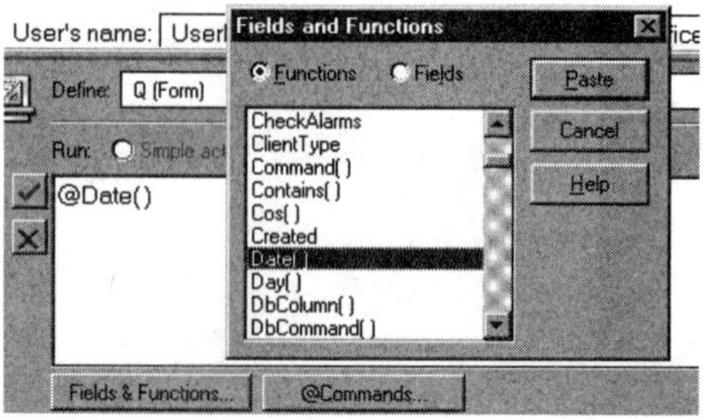

Figure 3.18 The Fields and Functions dialog box.

Table 3.2 Example @ Functions

@Function and Syntax	Example	Notes
@If (condition1; action1; condition2; action2; else action)	@If (Car = "Miata"; "Mazda"; Car = "Z3"; "BMW"; "This car is not defined.") If the field Car contains Tercel, this formula returns the text string "This car is not defined."	The @If statement evaluates the given conditions and then returns the specified action. If none of the statements apply, then the negative (else) action is returned. @If statements will always have an odd number of arguments to allow the negative statement. @If statements can be nested. You may use up to 99 arguments in an @If statement.
@ProperCase (string)	@ProperCase (UserName) or @ProperCase ("HOUSTON"), which returns "Houston"	@ProperCase places the argument (either a field, the result of another formula, or a text string) into proper sentence case. Similar @Functions include @UpperCase and @LowerCase.
@Trim (string)	@Trim (CarName) or @Trim ("Libby Schwarz"), which would return Libby Schwarz.	@Trim does not remove spaces internal to words, only those before and after.

continues

Table 3.2 Continued

@Function and Syntax	Example	Notes
@Text(argument)	@Text (Total), where the field Total is a numeric value, converts Total to a text value.	@Text is used to convert other types of constants or variables to text. This is used because only one type of operand or variable can be used in a formula. Use the Notes Help database to examine the time-date components of @Text.
@Created		Displays the creation date of a document
@Adjust (time-date; year; month; day; hour; minute; second)	@Adjust ([08/10/98]; 3;0;-1;0;0;0) returns 11/10/97	Adjusts a time/date forward using positive arguments and backward using negative arguments; you must account for all elements of a time/data field.
@Max (number; number)	@Max (5:2:8; 3:9:2) returns 5, 9, 8 (compares 5 & 3, 2 & 9, and 8 & 2 and returns the larger value)	Compares two numbers or number lists and returns the larger value.

@Function and Syntax	Example	Notes
@Subset (list; number)	@Subset ("Brian;" "Libby;" "Mike;" "Scott;" 2) returns Brian; Libby @Subset ("Brian;" "Libby;" "Mike;" "Scott;" -2) returns Mike; Scott	@Subset returns the appropriate number of members of the list:
@Length (argument)	@Length ("Total Seminars") returns 14, the number of characters in the string.	@Length determines the number of characters in a text string.
@Left (string; *n*), where *n* is the number of characters to return.	@Left ("Total Seminars"; 5) returns Total.	Returns the appropriate number of characters, counting from the left.
@UserName		Returns the current server or user name.

@Commands

@Commands (At Commands) are a special type of function that automate the standard Notes menu choices, as well as some special Notes commands. They are often used in buttons, actions, and agents. For example, @Command ([FileSave]) saves the current document in the same way that choosing File . . . Save or pressing Ctrl+S would. There are two types of @Commands, @Command and @PostedCommand. The only difference between the two types of @Commands are the order in which they are evaluated. @Commands are evaluated as they are reached in the document, in sequence from top to bottom, left to right. @PostedCommands are evaluated in order after all other @Functions or @Commands have been evaluated. This is true regardless of their location in the form. In Lotus Notes V3, @Commands run at the end, after all other @Functions, similar to the way @PostedCommands run in Notes Release 4. Forms converted from V3, therefore, have their V3 @Commands automatically converted to @PostedCommands to preserve the order of evaluation.

The basic syntax for an @Command is:

```
@Command([CommandName]; parameters)
```

Not each @Command has an additional parameter, although some have multiple parameters. Any parameters that are included will be separated with a semicolon. @PostedCommands use the same commands and have the same syntax. For example:

```
@PostedCommand ([CommandName]; parameters)
```

NOTE

There is a complete reference to @Commands in the Notes Help database or in the official Lotus publication, *Lotus Notes Programmer's Guide, Part 2.* For more information, please refer to this database or documentation.

A good example of an @Command is a button that automatically creates a new document or saves the current one. To create a button that automatically creates a new answer to one of the questions in our sample database, the button would have the following formula: @Command ([Compose]; "Answers"). When a user clicks the button, a new document would be created using the form *Answers.*

NOTE

@Commands are not always available. When creating the formula, if you do not see an @Command button in the formula window, you cannot create an @Command in that context. The @Command dialog box is shown in Figure 3.19.

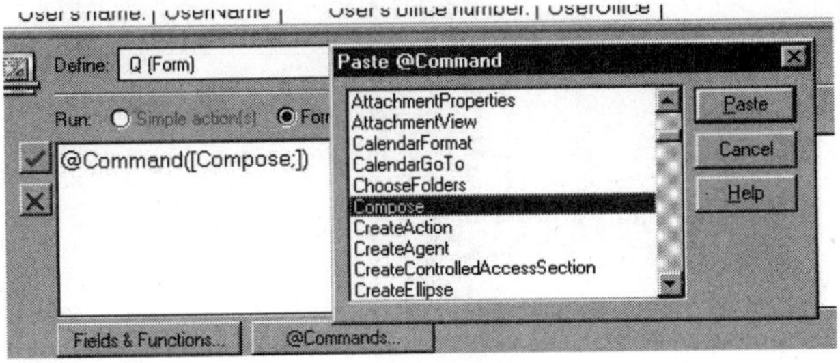

Figure 3.19 @Command dialog box.

Table 3.3 contains a few examples of @Commands. Again, for a more complete reference to all available @Commands, refer to the Notes Help Database or to the official documentation, *Lotus Notes Programmer's Guide, Part 2*.

Field Event Formulas

When we created fields earlier, we discussed the idea that in an editable field a designer could create a default value using the Default Value Event. Similarly, when you created a Window Title for your form in Chapter 2, you used a Window Title Event. These events, and the other field events, are good examples of a time when formulas can be used. The three main field events available for editable fields are Default Value, Input Translation, and Input Validation, as shown in Figure 3.20. The other field elements are outside the scope of the Application Development I and II exams. There are no field events for computed fields.

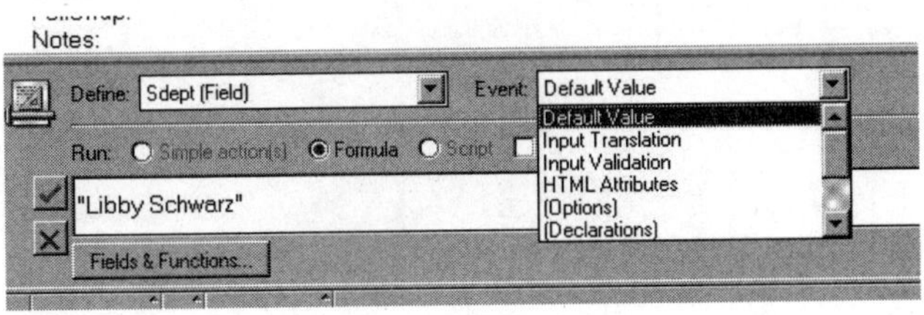

Figure 3.20 Field events.

Table 3.3 Example @Commands

@Command	Function	Example
[Compose]	Creates a new document with the specified form.	@Command ([Compose]; "Main Topic")
[FileCloseWindow]	Closes the active Notes window.	@Command ([FileCloseWindow])
[FileExit]	Closes all open Notes windows and exits Notes.	@Command ([FileExit])
[FilePrint]	Prints the current document or view.	@Command([FilePrint])
[ViewRefreshFields]	Refreshes the fields on the current document or the documents in the current view.	@Command([ViewRefreshFields])
[OpenNavigator]	Opens the specified Navigator in the current database.	@Command([OpenNavigator]; "Main Navigator')

A *Default Value* formula, as we have discussed already, places a value in an editable field that the user can accept or change. If there is an error in a Default Value formula, the form will not allow any input.

An *Input Translation* formula evaluates the data the user has typed into the field and then edits it. For example, the @Trim and @ProperCase functions described above would be placed in the Input Translation event to have the desired effect. These formulas are evaluated when the document is refreshed or saved. One option that we discussed earlier was the capability to refresh fields automatically as the user moved to the next field. If this is the case, the user will see the results of an Input Translation formula as they tab to the next field.

Input Validation formulas are used to require data to be entered into certain fields or to require a certain type of data to be entered into a field. This type of formula would be used to make a particular field, such as a Name or Categories field, required before the document could be saved. This type of formula, similar to the Input Translation formula, will be evaluated when the document is refreshed or saved, or, if the designer is using the Automatically Refresh Fields option (as discussed), the formula will be evaluated as soon as the user moves to another field.

Input Validation formulas are written using an @If function in combination with an @Success and @Failure function. For example, the following formula requires the Categories field to have a value before the document can be saved: @If (Categories = "";@Failure ("You must choose a Category before saving the document."); @Success). If the user exits without completing this field, he or she will receive an error like the one in Figure 3.21.

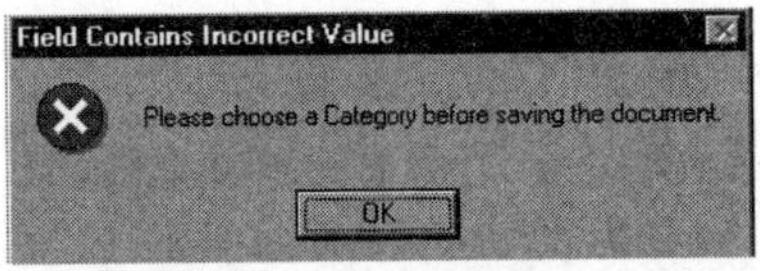

Figure 3.21 Input validation error.

EXAM TIP

Probably the most important @Function for the exam is the @If function. Know how to write, evaluate, and troubleshoot this @Function.

Formulas are used throughout Notes applications in both forms and views, as well as in other design elements of the database. This section has introduced you to the basic elements and syntax of a Notes formula.

Agents

Users and developers can create agents in databases to automate manual functions. An agent is similar to a macro in other applications, as it will automate a set of manual tasks. Unlike macros, however, agents are not recorded by doing the actions, but by using the Agent design window. The agent design window is shown in Figure 3.22. Agents are defined by the following elements: what triggers the agent to run, whether the agent is personal or shared, and how documents are selected for the agent to run on them.

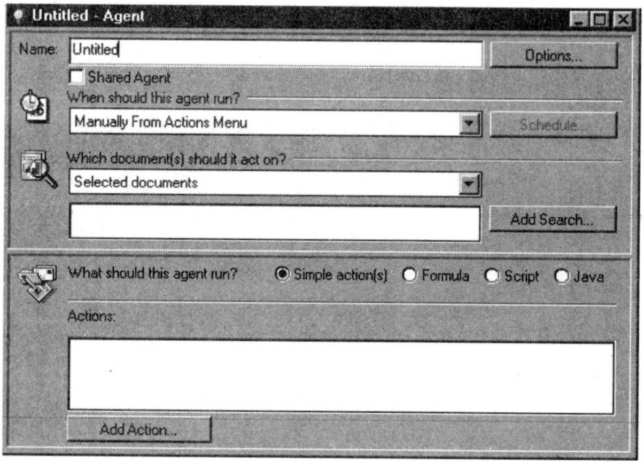

Figure 3.22 Agent design window.

To create an agent, choose Create . . . Agent from the menus. The Agent design window will be displayed, where you can name the agent. It is usually suggested to name the agent something descriptive, as it will appear on the Actions menu. You also have the opportunity to decide whether the agent is personal or shared. A shared agent can be created by someone with Designer level access or above. A personal agent, on the other hand, can be created by anyone who has been given the Create Personal Agents capability on the Access Control List dialog box. A shared agent can be used by other users of the database, whereas a personal agent can only be used by the person who created it.

NOTE

This is the only opportunity to decide if an agent is shared or personal. Once the agent has been saved, it cannot be changed from one type to another; it would have to be recreated to change this option.

The next option in the Agent design window is *When should this agent run?* This allows you to choose the trigger for the agent. Agents can be triggered either manually or automatically. The Application Development I exam only tests manually triggered agents. The options available from the drop-down list include the following:

- Manually from Actions list
- Manually from Agents list
- If New Mail has arrived
- If Documents have been created or modified
- If Documents have been pasted
- On Schedule Hourly, Daily, Weekly, or Monthly
- On Schedule Never

If an automatic trigger is selected, the designer can further designate the schedule of the agent using the Schedule . . . button. Agents can also be run from action buttons in forms and views, which will be discussed later in this chapter.

The next option in the Agent design window determines which documents the agent will act on. The *Which document(s) should it act on?* drop-down list has the following options for choosing documents:

- All documents in database
- All new and modified documents since last run
- All unread documents in view
- All documents in view
- Selected documents
- Run Once

In addition, a search query using forms, keywords, or other settings can be created using the Add Search . . . button. The Application Development II exam requires knowledge of the more advanced methods of selecting documents.

Finally, you have to define the action that the agent will perform. This can be a simple action, such as Modify Field or Move to Folder. The agent can also run a formula, based on the formula language discussed in this chapter. The agent could also run a LotusScript program, but that is beyond the scope of this book.

An example of an agent that can be created with a simple action is one that will modify a field value. To create an agent to modify a field value, select Create . . . Agent from the menu. Type a name for the agent and choose whether or not it will be shared. Ensure the agent is set to run manually from the Actions menu on selected documents. By clicking Selected documents, you require users to run the agent from a view or from within a document. Click the Add Action button and choose the Modify field action from the drop-down list. Select a field from the drop-down list as well. The field drop-down list displays all the available fields in the current database. Place a value in the value field and decide whether the value should replace the current value or be appended to it. If you modify a keyword field that uses synonyms, use the synonym as the value. You might have an approval field with the keywords yes and no, for example, with 0 and 1 as the synonyms. If you created an agent called Approve, it would replace the value of the Approve field with *0* for *yes*.

Actions

Actions can be associated with forms or views (see Chapter 4 for more information on View actions). They are available either from the Actions menu, similarly to agents, or from a button on the Action bar. To view the default Actions provided with Notes, choose View . . . Action Pane from the menu. The Action pane will be displayed, as shown in Figure 3.23, showing the default actions including Categorize, Forward, and Edit Document. These default actions are available from the Actions menu and can be added to an action button. To create a custom action for the form, choose Create . . . Action from the menus. This displays the Action properties InfoBox, shown in Figure 3.24.

The Action must first be named; as usual, choose a unique, descriptive name for the action, as it needs to be distinguishable to be chosen from a menu. You then have the option to choose a button icon from a predefined selection. Choose whether the action will be selected from the Actions menu, button bar, or both using the checkboxes on the InfoBox. The last option, position, determines where the button will be displayed on the button bar, or where the Action will be listed on the Actions menu.

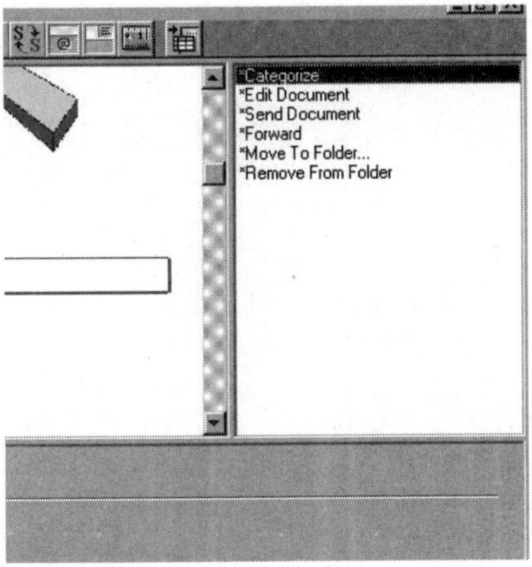

Figure 3.23 Action pane.

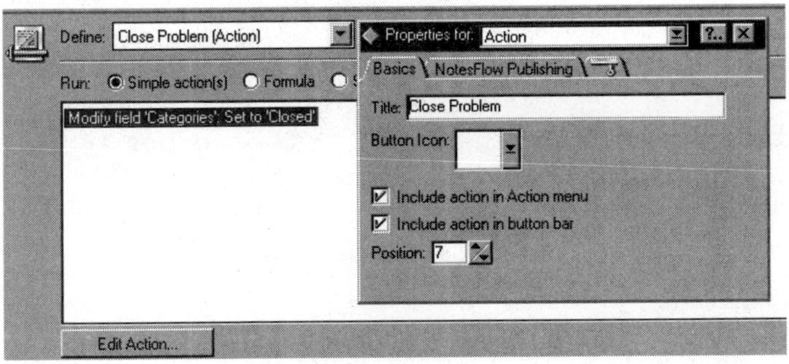

Figure 3.24 Action properties InfoBox.

There is a limited amount of space available on the button bar. It is often a good idea to use the Hide-When options available from the Action properties InfoBox, therefore, to show only the appropriate buttons, either using document events or a Hide-When formula. For example, an action to place a document in Edit mode would only be necessary when the document was opened for reading or previewing. Similarly, a common action in forms is to save and close the current document. This action would use two

@Commands and be available when the document was being edited. The formula for the action would be:

```
@Command ([FileSave]); @Command ([FileCloseWindow]).
```

Sample Database

Now that you have a thorough understanding of Notes fields and formulas, it is time to apply that knowledge to the sample database. The basic requirements for the two forms are listed below. Feel free to embellish on these few requirements until you feel comfortable. Again, the requirements for the database are limited to provide a review of the Application Development I required competencies.

Question Form Requirements

Following the static text you added in the previous chapter, add fields for each question. Suggested fields, including type and formula, are listed in Table 3.4. Let the user select their name from their address book. Give the user a default value for the name field that they can change if necessary. The Name field should be required. The OfficeNum must be exactly four digits long. Give the user the ability to pick his or her department from a list. The date of the question should be computed and displayed for the user to provide a date stamp. The questions will be shown in a view, categorized by *How do I* or *Something is broken* type questions. Users must select one of these two categories. The *How Do I* and *Something is Broken* static texts should only be visible as appropriate for the category. The user should be able to include graphics or other formatting in the description of the problem. There should be a button available when the form is opened for editing that saves and closes the form. Give the button the following formula:

```
@Command([FileSave]);
@Command([FileCloseWindow])
```

Note the ever-important semicolon!

The Questions form should look something like the figure on the next page when opened for editing.

Table 3.4 Suggested Sample Database fields in Question Database

Field Name	Field Type	Formula or Options	Notes
Name	Editable Names	Default Value: @Author Input Translation: @ProperCase(@Trim (UserName)) Input Validation: @If (Name = "";@Failure ("You must enter your name!"); @Success)	Name of the person asking the question. Choose the Use Address Dialog for choices option.
OfficeNum	Editable Number	Input Validation: @If (@Length(@Text(OfficeNum) != 4; "";@Failure("Office Number is invalid"); @Success)	Office number of the person asking the question.
Phone	Editable Number	Make a text pop-up hotspot: "Only enter your four-digit extension!"	Phone number of the person asking the question.
UserName	Editable Text	None	Name of the person needing the help. May be different from the person asking the question.
UserOffice	Editable Number	Input Validation: @If (@Length(@Text(OfficeNum) != 4; "";@Failure("Office Number is invalid"); @Success)	Office number of the person needing the help.

continues

Table 3.4 Continued

<table>
<tr><th>Field Name</th><th>Field Type</th><th>Formula or Options</th><th>Notes</th></tr>
<tr><td>UserPhone</td><td>Editable Number</td><td>Allow Multi-values
Make a text pop-up hotspot "Only enter the four-digit extension!"</td><td>Phone number of the person needing the help.</td></tr>
<tr><td>Dept</td><td>Editable Keywords</td><td>Create checkbox with the following choices: Accounting, Design, Human Resources, Manufacturing, Research, Sales, Support, Training. Make it four columns wide.</td><td></td></tr>
<tr><td>ProbDate</td><td>Compute when composed
Date</td><td>Value: @Date(@Now))</td><td>Automatic date stamp.</td></tr>
<tr><td>Categories</td><td>Editable Keywords</td><td>Create radio button—two options with synonyms: How do I | 1 and Broken | 2</td><td>This will help get the problem to the right person.</td></tr>
<tr><td>Describe</td><td>Editable Text</td><td></td><td></td></tr>
<tr><td>Now</td><td>Computed for Display
Date</td><td>Value: @Date(@Now))</td><td>Use in subform</td></tr>
</table>

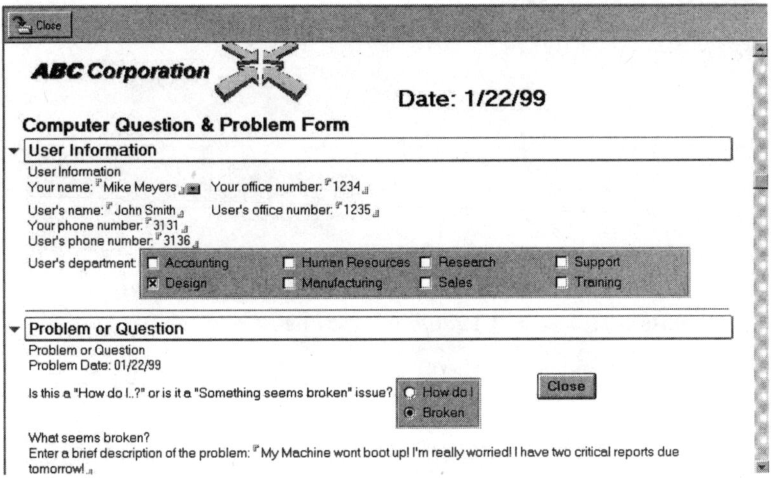

Figure 3.25 Questions form sample.

Answer Form Requirements

Again, following the static text you added in the previous chapter, you will add fields in the appropriate places on the form. The suggest fields for the Answers form are included in Table 3.5. The Answer form should have a section at the top that inherits information from the Question form to describe the question and the name of the person who asked the question. There should be fields that show the name and the department of the person responding to the question. There should also be three fields that are only visible when a document created with the Answer form is opened for editing. There should be an action bar button available when the form is opened for editing to Save and Close the form.

The Answers form should look something like this when opened for editing.

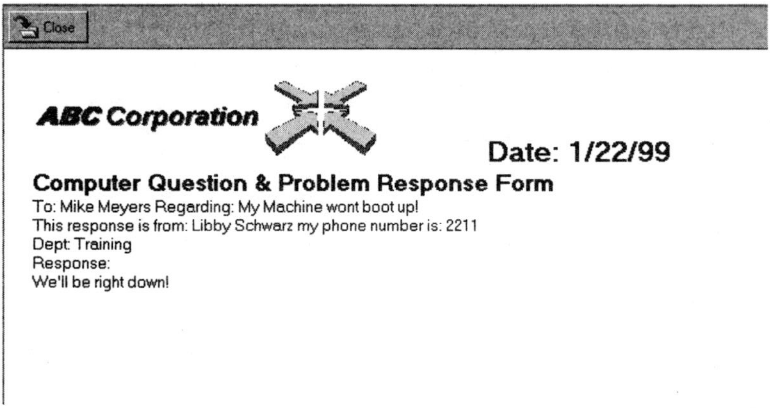

Figure 3.26 Answers form sample.

NOTE

Hint: To test your form as you build it, use Design . . . Test Form. Or, in Notes R4.6, choose Design . . . Preview in Notes.

Table 3.5 Suggested Sample Database fields in Answers Database

<table>
<tr><th>Field Name</th><th>Field Type</th><th>Formula or Options</th><th>Notes</th></tr>
<tr><td>UName</td><td>ComputedText</td><td>Value: Name
Make sure form has the "Formulas inherit values from selected document" option checked.</td><td>Name of the person asking the question.</td></tr>
<tr><td>Uquestion</td><td>ComputedText</td><td>Value: @left(Describe);60)
Make sure form has the "Formulas inherit values from selected document" option checked.</td><td>First 60 characters of problem description.</td></tr>
<tr><td>Support</td><td>Editable Text</td><td></td><td>Name of the person responding to the question.</td></tr>
<tr><td>Sphone</td><td>Editable Number</td><td>Allow multi-value
Make a text pop-up hotspot:
"Only enter your four-digit extension!"</td><td>Phone number of the person responding to the question.</td></tr>
<tr><td>Sdept</td><td>Editable Keywords</td><td>Create radio button—two options with synonyms: Support | S and Training | T</td><td>The department of the responding person.</td></tr>
<tr><td>Response</td><td>Editable Rich Text</td><td></td><td></td></tr>
</table>

continues

Table 3.5 Suggested Sample Database fields in Answers Database

Field Name	Field Type	Formula or Options	Notes
Followup should	Editable Text	Hide-When previewed or opened for reading	Date to follow up on this problem. This information not be seen by the person asking the question.
Notes	Editable Text	Hide-When previewed or opened for reading.	Support notes on this problem. This information should not be seen by the person asking the question.
Status	Editable Text	Default Value: "Open" Hide-When previewed or opened for reading.	Status of problem. This information should not be seen by the person asking the question.
Now	Computed for Display Date	Value: @Date(@Now)	Use in subform.

Review Questions

1. Joe wants to require the users of his application to type in their UserName before completing and saving the Question form. Which choice below will accomplish this?

   a. In the Input Translation event, the formula:

      @If (UserName = ""; @Failure ("Type your UserName, please."); @Success)

   b. In the Input Validation event, the formula:

      @If (UserName = ""; @Failure ("Type your UserName, please."); @Success)

   c. In a computed field called UserName, the formula:

      @If (UserName = ""; @Failure ("Type your UserName, please."); @Success)

   d. In the Input Validation event, the formula:

      @If (UserName = ""; @Success (Type your UserName, please."); @Failure)

2. Marie wants the waiters using her restaurant's online ordering application to be able to choose an item from the menu by clicking a button and being shown a list. What field type can she use to do this?

   a. Text field

   b. ListBox field

   c. Keywords field

   d. MenuList field

3. After creating a Customer form to allow users to type in their customer information, Lisa decides that she wants to reuse that customer information on her Customer Order form, too. The customer information is captured in these three fields: CustomerName, CustomerNumber, and Customer Address. What is the best way for Lisa to accomplish this?

a. Delete the fields from the original form. Recreate them as shared fields that can be used on both forms.

b. Put the fields on both forms using the same names. The data is shared automatically.

c. Choose each field and open its Properties InfoBox. Choose Share this Field from the Options tab.

d. Choose each field and click the Design menu. Choose Share this Field from the menu.

4. The keyword field that Liz is creating needs to allow multiple values. Which options allow her to do this?

a. Create a Keywords field of the Checkbox type. It automatically allows multiple values.

b. Create a Keywords field of the Radio button type. It automatically allows multiple values.

c. Create a Keywords field of the Dialog list type. It automatically allows multiple values.

d. Create a Keywords field of the Dialog list type. Choose Allow multi-values from the Basics tab of the Field Properties InfoBox.

5. Dave creates a keyword list. The choices in the keyword list are book titles for the books available for his employees to check out. They are all four to five words in length. He creates synonyms for the keywords. How and why does he do this?

a. He separates each synonym from the original choice with a backslash. He uses synonyms to make writing formulas that call the keyword choices easier to write.

b. He separates each synonym from the original choice with a backslash. He uses synonyms to save space in his database.

c. He separates each synonym from the original choice with a pipe, or vertical bar symbol. He uses synonyms to save space in his database.

d. He separates each synonym from the original with a semicolon. He uses synonyms to make writing formulas that call the keyword choices easier to write.

6. Wendy creates a form that she wants users to be able to add tables, graphics, and formatting to when they complete it. She needs to add what type of field to allow this?

   a. Rich Text

   b. Text

   c. Format Enabled

   d. Editable

7. The new application developer in your department is creating a formula to complete a computed field. She is sure that she has used the correct formula, but she says that she can't get the formula syntax to work. What would you tell her about the following formula?

   "This document was created by" @UserName "on" @Created "."

   a. She needs to add the text and field names together by using plus signs (+).

   b. She needs to add the text and field names together by using semicolons (;).

   c. She needs to change the @Created to text by using @Text (@Created).

   d. She needs to enclose the field names in quotes.

8. The users of Mike's application tend to leave their Caps Lock key on all the time. The data needs to be displayed in proper case when the users read the database, however. How can Mike make sure the data is displayed correctly?

   a. Mike can use an @ChangeCase function in the Input Translation field event.

   b. Mike can use an @ChangeCase function in the Input Validation field event.

   c. Mike can use an @ProperCase function in the Input Validation field event.

   d. Mike can use an @ProperCase function in the Input Translation field event.

9. Becky creates a field to capture the date that a document was created. In a computed field, she puts the following formula: @Created. Does this field accomplish her goal?

   a. Yes. The field always displays the date when the document was first created.

   b. No. The field needs to be a Computed for Display type of field or it will change every time the document is saved or refreshed.

   c. No. The field needs to be a Computed when Composed type of field or it will change every time the document is saved or refreshed.

   d. No. The field needs to be editable with the @Created as a default value or it will change every time the document is saved or refreshed.

10. Sue wants to place a button on her form that the user can click to Save and Close the document. After she creates the button, what formula can she use to accomplish this?

    a. @Command ([CloseandSave])

    b. @Command ([FileCloseWindow]); @Command ([FileSave])

    c. @Command (CloseandSave)

    d. @Command (FileCloseWindow) + @Command (FileSave)

11. Name three of the five defining characteristics of fields.

12. Pascale is creating a field that will be used later to group documents by category. What should she name this field? What data type should it be?

13. What is the correct evaluation of the following formula, assuming that the field Car is set to BMW: @If (Car = "BMW"; "You have a very nice car."; "VWBug"; "You drive a fun car."; "Bike"; "Someday you'll get a car."; "You mean you walk everywhere?").

14. What is the correct evaluation of the following formula, if the field Sandwich is set to "Peanut Butter and Jelly": @Length (Sandwich)?

15. Shelly wants to create an action button that will let users create a document with the Recipe form. What formula does she need to create for this action button?

16. Marcus wants to create an action button that will update the Status field on his Vacation Request form in the Human Resources database. The Status field is a keyword field with the following values:

    Approved |1

    Rejected | 2

    Delayed | 3

    Awaiting Decision | 4

    There is a default value formula in the keyword field that sets the value of the field to Awaiting Decision. The action button is called Approve and should set the field to Approved. How does Marcus create this action?

17. Tom wants to show an action button that places a document into edit mode on documents that are opened for reading and another action button that saves and closes a document that is opened for editing. Tom only wants the action buttons to show at the correct times. What does he need to do to create these and show them at the correct times?

18. Kathy needs to automate the approval process for a database she is creating; however, she does not want to create action buttons, as she wants the Approve capability in both forms and views in the database. What can Kathy create to automate the process of setting the Status field on the documents to Approved (1)?

Answers

1. b. In the Input Validation event, the formula:

   @If (UserName = ""; @Failure ("Type your UserName, please."); @Success)

2. c. Keywords field
3. d. Choose each field and click the Design menu. Choose Share this Field from the menu.
4. Both a and d
5. c. He separates each synonym from the original choice with a pipe, or vertical bar symbol. He uses synonyms to save space in his database. (It is also true that an advantage to using synonyms is that it makes writing formulas that contain the keyword choices easier.)
6. a. Rich Text
7. a and c. She needs to concatenate or add the text and field names together by using plus signs (+), and she needs to change the @Created to text data type instead of date data type by using @Text(@Created).
8. d. Mike can use an @ProperCase function in the Input Translation field event.
9. c. No. The field needs to be a Computed when Composed type of field or it will change every time the document is saved or refreshed.
10. b. @Command ([FileCloseWindow]); @Command ([FileSave])
11. The five defining characteristics are name, data type, computed or editable, how it is displayed, and whether it contains formulas.
12. The field should be named Categories. The field would most often be a keywords field.
13. "You have a very nice car."
14. 23—include all the characters and spaces of the value of the field or of a text string.
15. @Command ([Compose]; "Recipe")
16. Choose Create . . . Action. Use a Simple Action and click the Add Action button. Figure 3.27 shows the values in the Add Action dialog box. Notice that we used numbers instead of words because the keywords used synonyms.

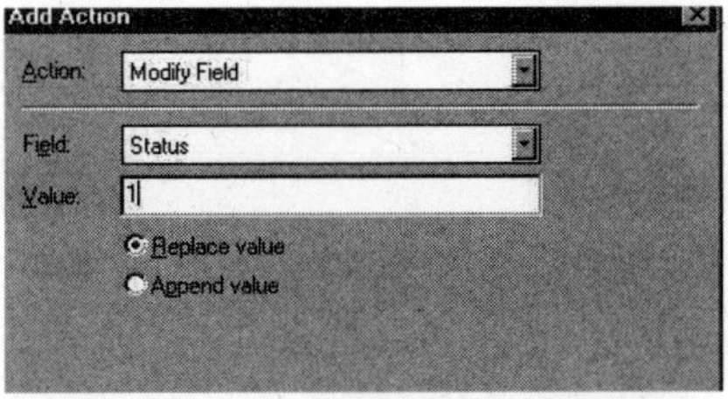

Figure 3.27 Add Action dialog box.

17. Tom creates two actions on the form using Create . . . Action. Use @Command ([EditDocument]), @Command ([FileSave]), and @Command ([FileCloseWindow]) as the formulas for the buttons. Use Hide-When options to hide the Edit button or the Save and Close button when the document is opened or previewed for editing.
18. Kathy can create an agent that runs the simple action of Modify fields. She can create an agent that will run manually on selected documents. This way the agent can be run either from within a document or on selected documents in a view.

CHAPTER 4

Views

Adding data to a form and then saving the form creates documents. At times, we may want to look at some part of the documents to query the database. This is performed using *Views*. A view is a list of the documents in the Notes database. Without a view, you cannot access the data that users have added to the database. Views can show all documents in a database, or specific documents, based on a selection formula.

Views in Notes databases compare somewhat to queries in relational databases. In Notes databases, however, views are predefined queries rather than ad hoc queries. The designer of a database creates the views, designs the format of the views, determines how the views will be organized, and chooses the documents that will be displayed. Another Notes design element that displays lists of documents is called a *folder*. The Application Development I exam covers your ability to create and modify views and folders in great detail, making the understanding of views and folders critical. In this chapter, we will learn how to create and modify views and folders, understand the different types of views and folders, and see how these elements are used within the realm of Notes.

Chapter Objectives

You should be able to answer questions based on the following objectives after reading this chapter:

- Create different types of views.
- Create folders.
- Add documents to a view.
- Add columns to a view.
- Use categories.
- Sort a view.

View Basics

Views are organized into two main elements: rows and columns. Documents in the database are organized into rows. A field or other type of information is represented by a column.

Folders are similar to views in design and purpose. The main difference between the two is the manner in which documents are selected for them. For a view, documents are selected through a selection formula created by the designer. In a folder, documents are selected and dragged or copied into the folder by a user. When a document appears in a folder, it is not removed from the view from which it was copied. However, if a document is deleted from a folder, it is deleted from the entire database. Both views and folders are displayed in the Navigation pane on the left side of a Notes database.

A designer must understand what is important about the documents to know what to show in a view. Decide on the purpose of the view before creating it. Knowing its purpose helps you decide which type of view to create, determine which properties to give the view, and understand how to design and define the columns in the view. The following sections describe these elements of creating a view.

View Types

Three main types of views are available in Notes and are described in the following list. View types can only be selected at creation time. After a view has been designated private or shared, it cannot be changed. The designer would have to create a new view to change this property of the view.

- *Shared views.* This type of view is created by the designer for many users. Most often, the shared view is displayed in the Navigation pane and in the View menu for easy access. Views can also be hidden, however. The shared view is also stored in the database. Users must have editor, designer, or manager access to the database to create shared views.
- *Private views.* Private views are created and used by only one user. Users with reader access or above can create private views. The storage for a private view is determined by the Create Personal Folders/Views checkbox in the ACL dialog box. Although this type of view is usually stored in the workstation's `desktop.dsk` file, if this checkbox is enabled, the view will be stored in the database.
- *Shared, personal on first use views.* This type of view enables the designer to create views for the users that are shared until the first time the user opens the view. At this time, the view becomes a private view and is stored based on the Create Personal Folders/Views checkbox. One thing to note for the exam is that the view will no longer be updated with changes after it has become private. To see any changes after opening the view, the user will have to delete the private version of the view and open shared version again.

NOTE

Note that *shared, personal on first use* views are not security measures. Even if you use a selection formula to omit certain documents from the view, a user can create a private view that shows all documents in the database.

Similarly, there are both *shared, private* and *shared, personal on first use* folders. These folders have the same elements and characteristics of the views. Again, the main difference between views and folders is the manner in which documents are selected for them. Views use selection formulas, and users move or copy documents into folders.

Create a View

A view (or a folder) can be created in a variety of ways, similar to a form. You can copy a view from the current database or another database. Alternatively, you can create a view from scratch. To copy a view from the current database, you create a new view and inherit the design features. The steps for creating a new view are listed here:

1. Choose Create . . . View . . . from the menus, or if you are in a Design view, click the Design View SmartIcon. This displays the Create View dialog box, displayed in Figure 4.1.

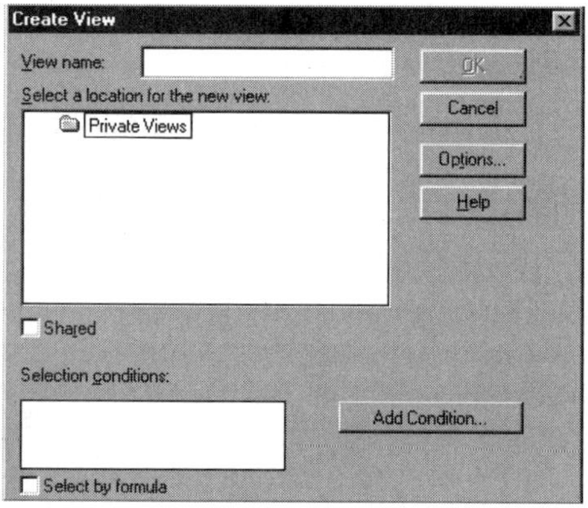

Figure 4.1 Create View dialog box.

2. Give the view a descriptive name. The view can also be given an alias (similar to a form name's synonym) to make changing the name or using the view name in formulas easier. The name is case-sensitive and can include any characters, including letters, numbers, and spaces. The view names can also be cascaded to organize the View menu for the users. To create an alias (synonym) in this dialog box, use the | symbol (pipe or vertical bar). To cascade names from the View menu, use the \ symbol (backslash).
3. Select whether the view is *Shared, Private* or *Shared, Personal on first use* using the Shared checkbox.
4. Select a location for the view by clicking the folder or view where you want to store this view.
5. To select which documents will be displayed in this view, use a selection formula. The selection can be created in this dialog box, either by using simple selection conditions created in the Search Builder window or by creating a custom selection formula. To open the Search Builder window, shown in Figure 4.2, click the

Add Condition . . . button. Then add the appropriate condition based on author, form, form used, field, or date. For example, you could select all documents based on the author for a shared, personal, on-first-use view. Alternatively, you could select a view based on the form used. Custom selection formulas are described later in this chapter and in Chapter 8.

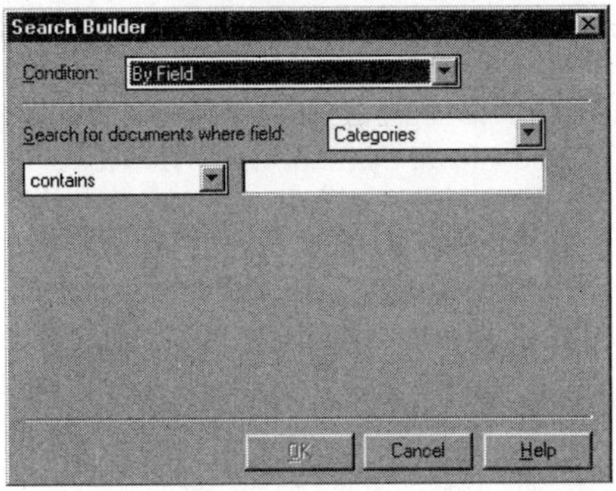

Figure 4.2 Search Builder window.

6. Additional options are available by clicking the Options button. This displays the Options dialog box, shown in Figure 4.3, which enables the user to inherit the design from another view. To inherit the design for the new view from an existing view, click the name of the already created view in the Inherit design from list box. To create a view from scratch, click the Blank choice in this list box. To go immediately to the design for the view, click the Design now checkbox.

Figure 4.3 Options dialog box.

7. Click OK to create the view. If you chose Design now, the view automatically opens in design mode. If not, you will be able to open the view from the Design . . . Views view to begin editing the view.

NOTE
You will only have one opportunity to decide whether a view or folder should be Shared, Private, or Shared, Personal on first use. This opportunity occurs only when you create the view or folder.

Creating Folders is essentially the same as creating Views. Choose Create . . . Folder. Give the Folder a name and choose whether it is *Shared, Private* or *Shared, Personal on first use.* Click the Options button to inherit the design of a particular view or to design the folder. There are no selection options available on a Create Folder dialog box.

View Properties InfoBox

The View is further defined by the View Properties InfoBox. This dialog box enables you to set the default options for the Views you create. Expect to see a variety of questions based on the View Properties InfoBox on the Application Development I exam. Each tab and option on the View Properties InfoBox is described in the following paragraphs.

The Basics Tab

The view can be named (or renamed) using the Basics tab. Although the View may already have been named when created, this tab, as shown in Figure 4.4, enables the developer to rename the view, to create an alias (synonym) for the view, and to create a comment about the view. The options for the Basics tab are described here.

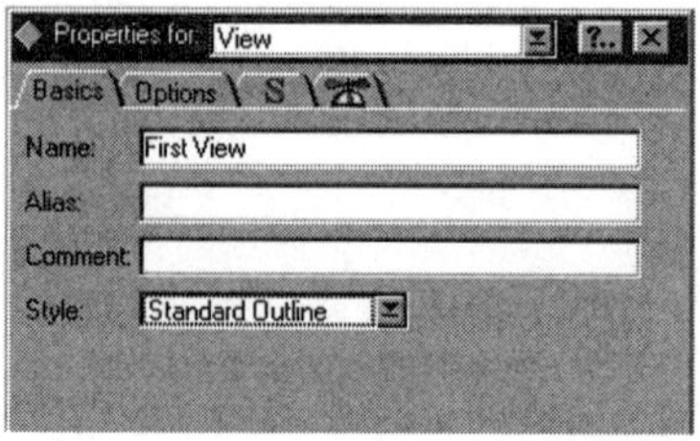

Figure 4.4 View Properties, Basics tab.

- *Name.* The Name field enables the designer to give the view a name. The name is case-sensitive and can include any characters including letters, numbers, and spaces. To display cascading names from the View menu, use the \ (backslash) symbol, similarly to the way cascading form names were created earlier.
- *Alias.* In the case of views, synonyms are called Aliases. In addition, there is a specific field for this purpose, as shown here, although an alias can also be created using the | (pipe or vertical bar) symbol in the name field. The alias, like a form synonym, enables the name of the view to be translated or changed without changing any formulas that may use the view name.
- *Comment.* This field enables the designer to comment on the view or purpose of the view.
- *Style* (R4.5X and above). This enables the designer to choose between standard design and a calendar style design.

The Options Tab

The Options tab on the View Properties InfoBox, shown in Figure 4.5, enables the developer to set some of the default attributes for the view. These attributes are described in the following paragraphs.

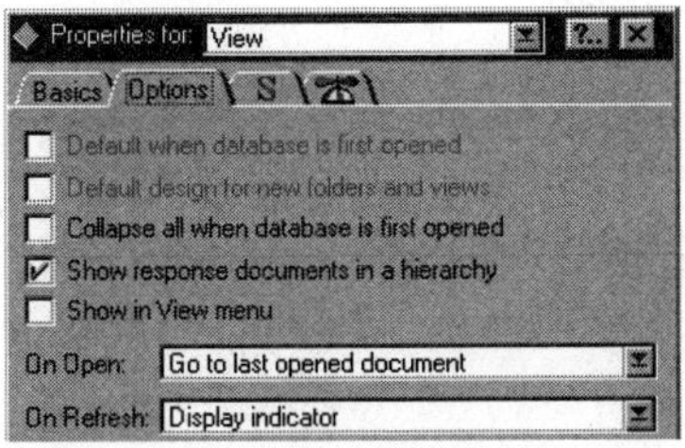

Figure 4.5 View Properties, Options tab.

- *Default when database is first opened.* If this option is selected, this view will automatically be the default when the database is opened. This option will be ignored, however, if the database launch property (set in the Database Properties InfoBox) defines a navigator with a particular view. Only one default view is allowed per database. The default view will be indicated in the view list with an asterisk (*).

- *Default design for new views and folders*. If this option is selected, the current view will be the default design for any new views or folders created in this database. In other words, new views and folders will automatically inherit design elements from this view. The designer can choose to inherit design information from another view by clicking the Options button when creating a new view.
- *Collapse all when database is first opened*. If documents are grouped using categories in the view, this option forces all the categories to be collapsed, hiding the documents, when the database is first opened. Categories are discussed in greater depth later in this chapter.
- *Show response documents in a hierarchy*. This option is one of the requirements necessary to show the response type and response-to-response type documents in the view indented under their parent documents.
- *Show in view menu*. This option determines whether the View is listed in the View menu. In R4, the Navigator pane is often used instead of the View menu.
- *On open*. The On Open options determine what is automatically displayed when the view opens. This setting can affect how quickly the view is opened when selected by a user, especially in the case of databases with a large number of documents. These options include the following:
  - *Go to last opened document*. When the view is opened, the last document that the user had opened is displayed in the view. This is the default setting.
  - *Go to top row*. This option displays the first document in the view each time that the view is opened.
  - *Go to bottom row*. This option displays the last document in the view each time that the view is opened.
- *On Refresh*. Each view displays the results of the view index. This index is an internal list that tracks the order and number of documents that should appear in this view. As documents are changed, deleted, and added, the view index changes. The view index must be refreshed to display the changes in the view. The speed of using the view can be affected by this setting. This option sets the view to refresh in one of the following ways:
  - *Display indicator*. With this option, a blue circular arrow refresh indicator is displayed in the upper left corner of the view when

the view needs to be refreshed. When the indicator is displayed, the user can either click the indicator or press F9 to update the view.

- *Refresh display*. This option forces the view to refresh itself automatically when changes are made.
- *Refresh display from top row*. This option forces the view to refresh automatically, beginning with the first (top) document in the view.
- *Refresh display from bottom row*. This option forces the view to refresh automatically, beginning with the last (bottom) document in the view.

The options to open the view and to refresh the display from the top row and bottom row should be chosen based on how the view is organized. If the view is being sorted based on a date column, for example, documents are added either to the top or to the bottom of the view when they are created. If the newest documents appear at the top first, open from the top of the view and then refresh from the top of the view, and vice versa.

The Style Tab

The Style tab, as shown in Figure 4.6, allows the developer to create the visual presentation of the view, including the background, columns and rows, and titles. The options available on this tab are described in the following paragraphs.

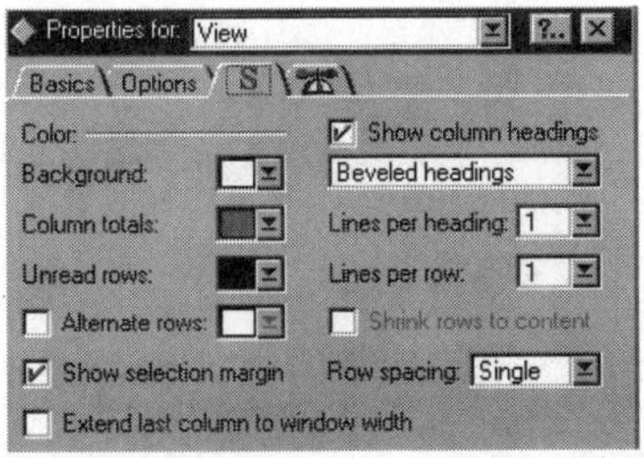

Figure 4.6 View Properties, Style tab.

- *Background*. This option allows the designer to choose a color for the background of the view. By default, the view background is white.

- *Column Totals*. This option enables the designer to create a separate color for the background of the columns that have the Total option set. The Total option, adds the values in a column and creates a total. By default, this option is displayed in dark gray.
- *Unread rows*. If the designer chooses to display unread marks, using the Advanced tab, this option enables those documents that are unread to be displayed in a different color. The color is black by default.
- *Alternate Rows*. This option enables the designer to display every other document in the view in a different background color. This option is disabled by default.
- *Show Selection Margin*. The margin to the left of the documents, which enables the user to select multiple documents is called the *selection margin*. This option enables the designer to determine whether this is displayed in the view. If this option is disabled, a user can still select multiple documents by using Shift+Click. When more than one document is selected, the Selection Margin is displayed automatically. This option is enabled by default.
- *Extend last column to window width*. This option forces the last column in the view to extend to the far right of the window, regardless of column or view widths.
- *Show column headings*. This option enables the designer to determine whether the column headings will be displayed in the view. By default, the column headings are displayed. There may be cases, however, where the information does not need a heading and this would not be desirable.
- *Heading Display Type*. When the Show column headings option is selected, the designer also has the option to determine whether the headings will be displayed as simple headings or beveled headings. Simple headings use the background color of the view to display the column headings; whereas; beveled headings use a gray three-dimensional effect to display the headings.
- *Lines per heading*. When the *show column headings* option is enabled, the designer has the ability to determine the number of lines available for the column heading. If the column headings are longer or consist of multiple words, this setting enables them to use more than one line. This allows more columns to fit in the view. This setting can be from one to five.
- *Lines per row*. This option enables each row to spread over multiple lines. In a case where an address or a large amount of text is

going to be displayed in the view, this allows the lines to take more than one line to display each document. This setting can be from one to nine.

- *Shrink Rows to content.* When the *Lines per Row* option is set to more than one, the designer has the ability to shrink rows to content. This allows those rows that do not need more than one line to display the data smaller, saving space on the screen. This ensures that only the necessary lines, up to the number specified in the Lines per Row option, are used for display.
- *Row spacing.* This option enables additional space to be inserted between rows for clarity. Especially in cases where long lines of text are being displayed in a view, having some extra space between the lines makes reading the view easier for users. The settings for this option can be single, 1 1/4, 1 1/2, 1 3/4, and double-spaced.

The Advanced Tab (The Propeller-Beanie Icon)

The Advanced tab, which looks like a beanie hat with a propeller on top, enables the designer to set additional options for the view. These options, as displayed in Figure 4.7, are related to indexing, unread marks, and the form used to display the documents in the view. These options are described in the following paragraphs.

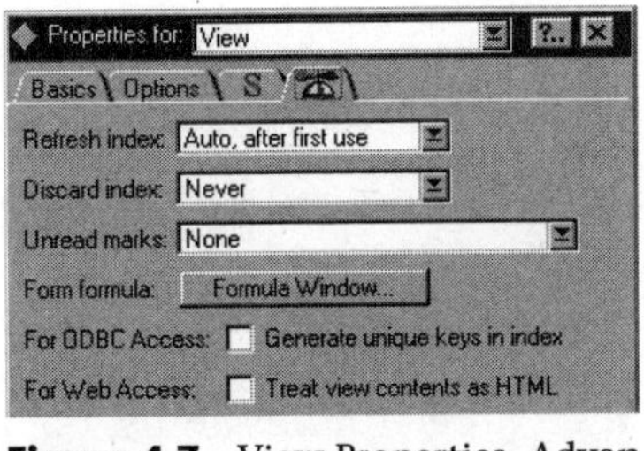

Figure 4.7 View Properties, Advanced tab.

- *Refresh index.* The view index, as described earlier, is the internal list that lets Notes determine which documents should be in each view. This option enables the designer to determine the refresh rate of the view index. The options for the index refresh rate are described in the following list:
  - *Auto, after first use.* This option forces the index to refresh every time the view is opened after the first time the view is opened.

This is the default option. Changes are added to the view index incrementally as they are made. This is easy for the users, as they never have to be concerned about the accuracy of the view. The views may take longer to open, however, as changes are written to the view index.

- *Automatic*. This option is similar to the default option, in that the index is refreshed automatically. The user does not have to open the view for the index to update, however, as in the default. This ensures both that changes are reflected in the views and that views open more quickly.
- *Manual*. This option forces the user to be responsible for updating the view. In the case of a large database that does not change regularly, this might be a good choice, as it allows the database to open more quickly.
- *Auto, at most every x hours*. This option is the compromise that enabless the view index to refresh automatically based on a schedule. If a large database changes often, this allows the database both to open quickly and to be updated on a regular basis. If the database has changed since the last update, the refresh icon will be displayed for the user to update the view manually.

- *Discard index*. This option determines when the view index for this view will be deleted. Whenever the index is deleted, the user will have to wait while a new index is generated, which can take several minutes. The reason for deleting a view index is usually related to disk space savings. The options for discarding the view index are described in the following paragraphs:
  - *Never*. If the designer selects this option, the view index is never deleted. Changes to the view index are made to the existing index. This ensures that the users do not have to wait for the view index to be created when they open the view. This option, however, uses more disk space than other options and should be selected for frequently used views.
  - *After each use*. This option deletes the view index each time the database is closed. This forces the user to wait for the view to rebuild each time the view is opened. Although this may save some diskspace, it should only be used for very infrequently used views.
  - *If inactive for x days*. This option enabless the designer to compromise between the speed and disk savings. This option allows the designer to set a number of days of inactivity, after

which the view index will be deleted. If the view index has been deleted, the user will have to wait while it is recreated when the view is opened.

NOTE

Users always have the ability to refresh the view by pressing F9 or clicking the refresh indicator. The refresh indicator appears anytime changes are made to the database but have not been updated in the view. The refresh indicator is a blue curved arrow on the left side of the column headings.

- *Unread marks*. On the Style tab, the designer chooses whether unread rows are displayed in a different color. This option enables the designer to determine whether unread marks will be tracked in the database. The options for unread marks are described in the following paragraphs:
  - *None*. This option, which is the default, does not show a special indicator for unread documents in the database. This option uses the least processor overhead, and therefore, opens the view the fastest. This option should be used if the read or unread status is not significant for the documents in this database. Even though Unread documents will not be visually marked, the Navigate Next or Previous Unread SmartIcons will allow the user to find unread documents in the database.
  - *Unread Documents only*. This option displays unread documents with an asterisk as well as any color options chosen by the designer on the Style tab. If the view is categorized, however, the unread marks will only appear when the category is expanded enough to display the unread documents in the view. This option is the compromise, as it opens faster than the standard option but slower than when unread documents are not marked at all.
  - *Standard (compute in hierarchy)*. This option always displays a special indicator to notify the user about unread documents. If the documents are in a collapsed category, the asterisk and color options are displayed on the category until it is expanded. This option uses the most processor overhead and is therefore slower to open, but it gives the users the most information about the documents. This option is especially useful in categorized views that have time-sensitive material or wherever the new documents are important to the users.

- *Form formula*. This option enables the designer to determine which form will be used to display documents opened from this view. The determination is made by Notes, using the following guidelines:
  - When a form is stored in a document, that form will always be used to display that document.
  - If no form is stored in the document, and a designer has used a form formula for the view, that formula determines the form used to display the documents in the view.
  - If no form is stored in the document and a designer has chosen not to use a form formula, the form used to create the document will be used to display the document, as long as it is available. If that form is not available, the default form for the database will be used to display the documents.
- *Unique Keys in index (for ODBC access)*. This tab assigns special keys for files that get their data from a source external to Notes, via the ODBC (Open Database Connectivity) standard.

The Security Tab

The Security tab determines which users and groups can use this view. The security elements of views are covered in Chapter 11, *Security in Applications*.

Selecting Documents for the View

When designing a view, a developer must determine which documents should appear in that view. When you create the view, you can choose the documents using the Search Builder. Additionally, either when creating the view or at any time, you can create a Selection formula to determine what documents to display in the view. The use of the Search Builder window to select documents for the view was discussed previously in this chapter. Now, let's discuss the use of custom selection formulas to select documents for the view.

To add a selection formula to a view, click the Define drop-down list at the top of the design pane. Choose View Selection from the list. If you choose the Easy radio button, use the Search Builder window by clicking the Add Condition button. If you choose the Formula radio button, you can create the selection formula in the formula window below. By default, the selection formula is SELECT @All, which displays all documents in the database. Another option for a selection formula includes selecting by the form or forms used, such as the following:

SELECT Form = "Answers" | Form = "Questions"

This selection formula would display any documents created with either the form Answers or the form Questions (see Figure 4.8).

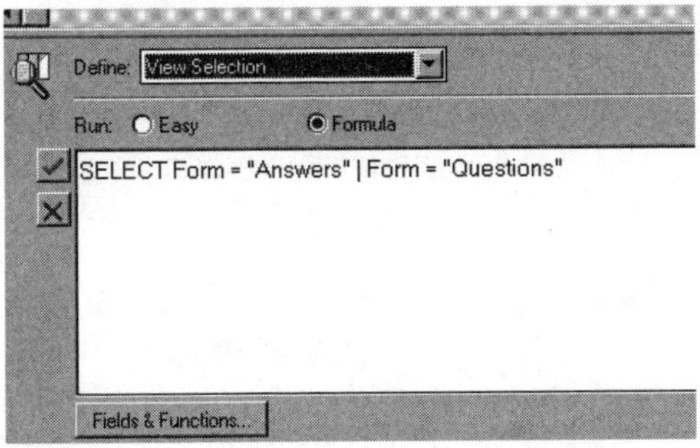

Figure 4.8 Creating a selection formula.

Another example:

SELECT Form = "Questions" & Status = "Unanswered"

This selection formula displays all documents created with the form Questions that have the value Unanswered in the Status field.

Also:

SELECT Form = "Questions" | @IsResponseDoc

This selection formula displays all documents created with the Questions form as well as all documents that are response documents. This would be useful when there are multiple main documents in a database.

A few @Functions are used primarily for selection formulas (or selective replication formulas). Some of these @Functions are discussed in the following paragraphs.

The @AllChildren function refers to all documents that are response documents to a particular main document. @AllChildren only refers to immediate responses of a document, not any response-to-response documents. This is different from the @IsResponseDoc selection formula above because the @IsResponseDoc includes any documents created with a response or response-to-response type of form. The @AllChildren function will return only documents that respond to other documents already selected. An example of this function:

SELECT Form = "Main Topic" | @AllChildren

This selection formula displays the documents created with the Main Topic form and any documents that are immediate responses (or children) to those documents.

The @AllDescendants function is similar to the @AllChildren but includes in a view the response-to-response documents that are the "grandchildren" to a main document. For example:

SELECT Form = "Main Topic" | @AllDescendants

This selection formula includes the documents created with the Main topic form, the documents that are responses to those documents, and the documents that are the responses-to-responses.

Figure 4.9 displays the differences between @IsResponseDoc, @AllChildren, and @AllDescendants.

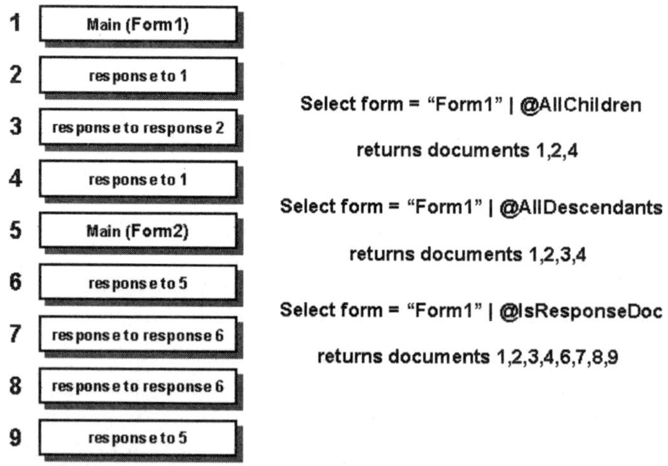

Figure 4.9 Differences between @IsResponseDoc, @AllChildren, and @AllDescendents.

Adding Columns to the View

After the view has been created, the appropriate documents have been selected, and the default attributes have been set, the designer must create and define columns for the view. The columns in the view determine what information the user sees about the documents that are displayed in the views. Columns can be defined by a simple function, a field, or a formula. (Folder columns are created in the same way as View columns.)

You have multiple options for creating a column in the view. First, the designer can double-click the gray area where the column headers are displayed. This automatically creates a column and displays the Column Properties InfoBox. Second, the designer can right-click the column header area and choose either to insert or to append a new column. The designer can also use the Create menu or the appropriate SmartIcon to add a new column.

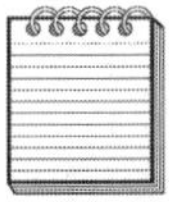

NOTE

Inserting a column places it to the left of the current one. Appending a column places it to the right of the current column. Double-clicking to add a column appends the column and using the Create Insert New Column SmartIcon inserts the column.

Column Properties InfoBox

After the column has been added to the view, the designer has to define the column attributes, using the Column Properties InfoBox, and has to determine the contents of the column, using either a function, field, or formula. First, we will discuss the attributes to be set in the Column Properties InfoBox.

THE BASICS TAB

The Basics tab of the Column Properties InfoBox, shown in Figure 4.10, enables the designer to determine name, width, and other general style attributes of the column. These options are discussed in the following paragraphs:

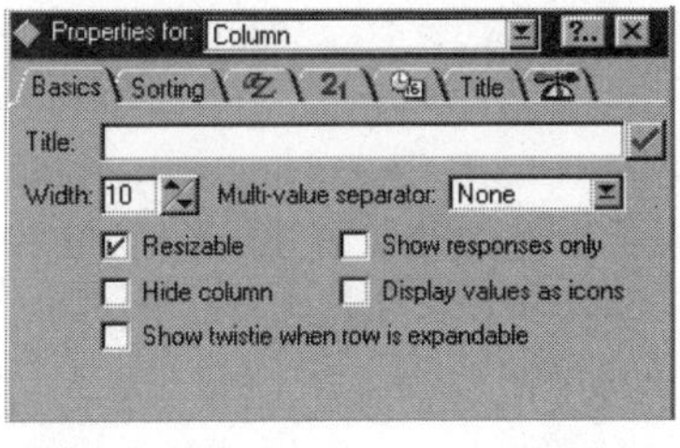

Figure 4.10 Column Properties, Basics tab.

- *Title*. The title field gives the designer the opportunity to give the column a descriptive title. The title should be kept to a reasonable length, as it will be cut off if the column width is not wide enough to display it all. Column titles are not required.
- *Width*. This option enables the designer to set the default width of the column. By default, it is set to 10. Change the width by dragging the column header or by changing this number.
- *Resizable*. By choosing this option, the designer gives the user the power to change the width of the column as they use the view. This is checked by default.
- *Hide Column*. This option hides the column from the view. This type of column can be used for sorting documents.
- *Show Twistie when row is expandable*. When the designer has chosen to categorize columns in the view, the categorized columns can be expanded or collapsed. This option shows the user when the category is collapsed and can be expanded, giving them a way to do this (clicking the twistie) other than using the Expand and Collapse SmartIcons.
- *Multi-value separator*. When a column will contain multiple values per document, such as in displaying keyword choices or names, this option enables the designer to determine how these values are displayed to the user. The default choice is None, which displays the data as it is stored internally. The designer can also choose to separate the multiple values with commas, spaces, semicolons, or new lines.
- *Show responses only*. This option is required to show only a response or a response-to-response type of document in a column. For example, most databases that include a discussion element will want to display the hierarchy of main document with its children and descendants indented underneath. An example of this type of view is available in the Discussion Thread View of a database created with the Discussion template, shown in Figure 4.11. The related option of Show Response Hierarchy (in the View Properties InfoBox) must also be selected to create this view effect.

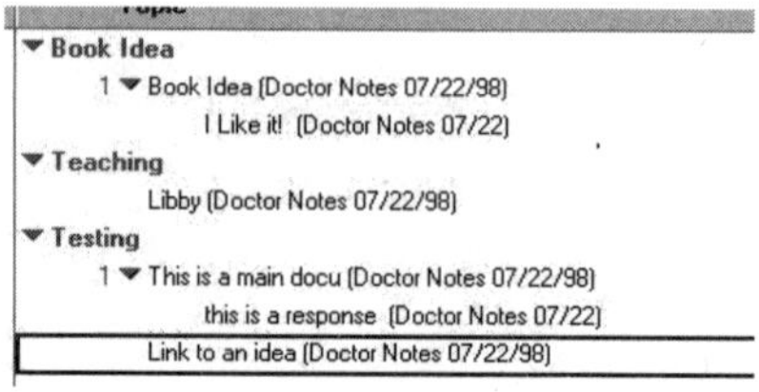

Figure 4.11 A typical Discussion database, showing Discussion Thread view.

- *Display values as icons.* Notes provides 170 icons, available in the Notes Application Developer's Guide, to be displayed in a column instead of text or numbers. This option is used in views to indicate the type of document and the presence of attachments. To display an icon, the column requires a formula that results in a number between 0 and 170. The available icons are shown in Figure 4.12.

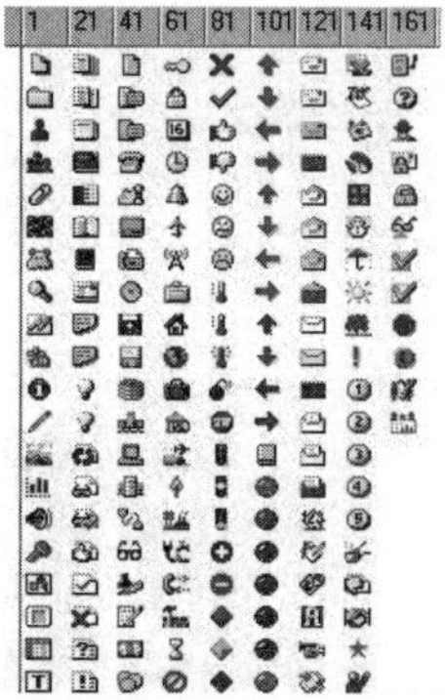

Figure 4.12 Available view icons.

THE SORTING TAB

The Sorting tab, shown in Figure 4.13, enables the designer to determine the order in which documents are displayed in the view and how those documents will be organized. The options available on this tab are discussed in the following paragraphs.

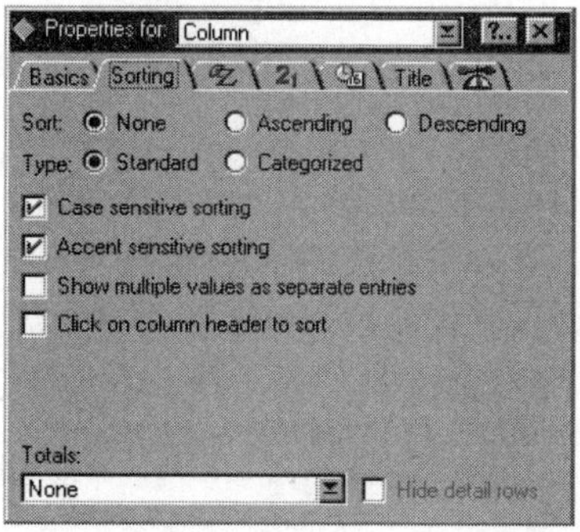

Figure 4.13 Column Properties, Sorting tab.

- *Sort.* The Sort option enables the designer to sort documents in ascending order, descending order, or not at all. By default, columns are not sorted. Notes sorts in the following order: numbers, letters, accented letters, and punctuation or special characters.
- *Type.* If the column is sorted, this option determines whether the sort type will be Standard or Categorized. The Categorized option, which is discussed in more detail later in this chapter, allows the designer to group documents together based on the values in a predefined field.
- *Case Sensitive Sorting.* By default, sorting is case-insensitive. This option gives the designer the ability to enable sorting based on case.
- *Accent Sensitive Sorting.* This option enables the designer to sort based on accented letters.
- *Show multiple values as separate entries.* Most fields in Notes (except Rich Text and a few keyword fields) can display multiple values in a single field. This option gives each of the multiple values its own entry in a sort.
- *Click Column Header to Sort.* This option gives the designer the ability to enable the users to sort the data in the view at any time. The options allow the user to sort ascending, descending, or both. These options add arrows to the column header that the user can click to sort the column. The list also contains a *Change to view* option. This option enables the designer to designate a view to be switched to when the user clicks the column header. This option can confuse users, so it is usually a good idea to describe what will happen in the column header, for example, *Click Here for the By Date View.*
- *Secondary Sort Column.* If the *Click on column header to sort* option has been enabled, this option gives the user an additional sorting column. This refines the sorting order without requiring any user action.
- *Totals.* This option enables the values in a column to be totaled. It is not tested on the Application Development I exam.

EXAM TIP

The remaining tabs on the Column Properties InfoBox are only tested in passing, if at all. Therefore, while they are described here, they are not discussed with the same level of detail as the Basics and Sorting Tabs, on which the Application Development I exam concentrates.

THE TEXT TAB

The Text tab, shown in Figure 4.14, gives the designer the opportunity to determine how the text in the columns will be formatted. After selecting the font, size, style, color, and justification, the designer can also click the Apply to All button to apply the formatting to all columns in the view.

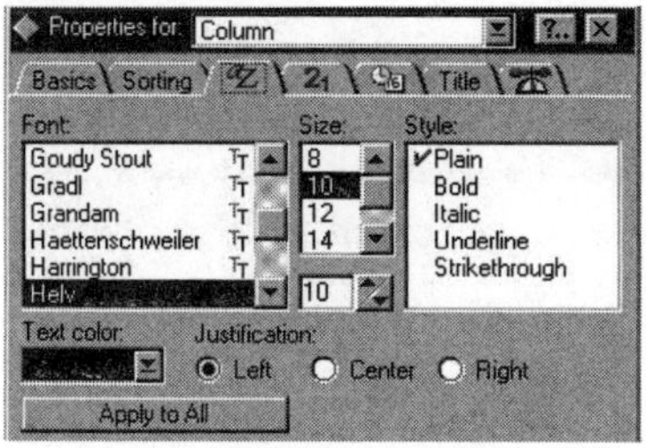

Figure 4.14 Column Properties, Text tab.

THE NUMBER TAB

The Number tab enables the designer to select the style and formatting of numbers used in the columns. This can include currency style for columns that display dollar amounts as well as parentheses for negative numbers and punctuation for larger numbers. The Number tab options are shown in Figure 4.15.

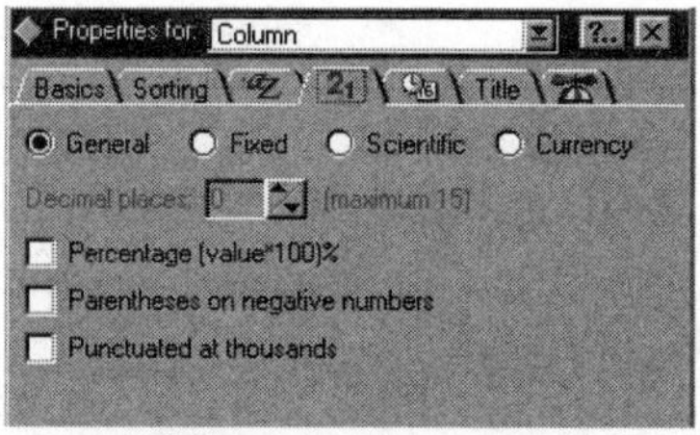

Figure 4.15 Column Properties, Number tab.

THE TIME/DATE TAB

The Time/Date tab enables the designer to format any date or time elements that may be included in the column. The options on this tab are similar to those already discussed for time fields.

THE TITLE TAB

The Title tab, which is similar to the Text tab, enables the designer to change the formatting of the column headers. Remember that if the column header is too long for the column, it will be truncated.

The Apply to All button on this tab allows the designer to set all the column titles in the view to the same formatting.

THE ADVANCED TAB

This tab is not discussed on the exam.

Define the Column

When the attributes for the column are set using the Column Properties InfoBox, the designer has to determine the value of the column. The value of a column can be determined in three ways: a simple function, a field, or a formula. These methods for creating a value in the column are described in the following paragraphs.

USING A FUNCTION

Using a Simple Function to determine the value of a column enables the designer to use values like the Author's simple name, Author's distinguished name, creation or modification date, size, or information about the document's attachments for the column value. To use a Simple Function for the column value, click the Simple Function radio button in the design pane for any column. The list of simple functions that are available are listed in the formula window, as shown in Figure 4.16. Choose the appropriate function. To preview the values of the columns, click the Refresh button. If there are any documents in the database that would be displayed in this view, they are used to preview the values of the columns. The simple function Number in View is the default value for all new columns.

USING A FIELD

Another common way to determine the value displayed in a column is to use a field value from the form. This takes the value of a field on the documents displayed in the view and displays it directly in the column. To use a field value to determine the display value of the column, click the Field radio button in the display field of the Design pane for the column. After choosing to use a field for the column value, choose the field that you want to use from the list presented in the design pane.

NOTE

You may be prompted with a dialog box like the one shown in Figure 4.17. When a column is changed from simple function to field or to formula, the previous existing value is lost. If you are unsure about losing a formula or function, copy the column before changing the value.

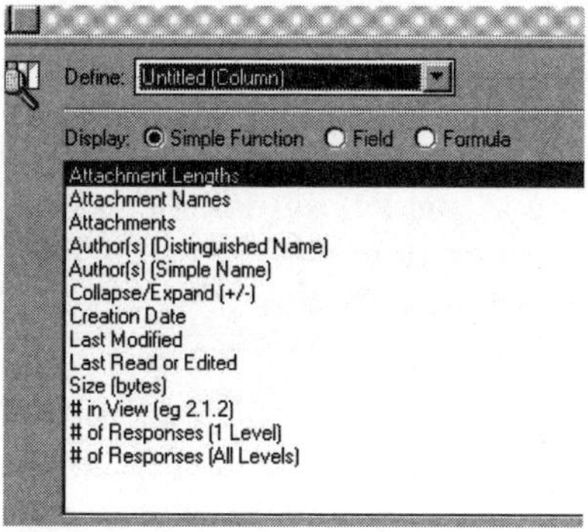

Figure 4.16 List of Simple Functions for column values.

NOTE
Certain field types cannot be used in column formulas. These include the following: Rich Text fields, computed for display fields, and any encrypted fields.

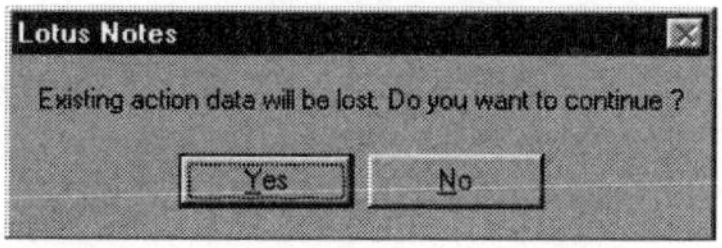

Figure 4.17 The Existing data will be lost dialog box.

USING A FORMULA TO DETERMINE COLUMN VALUES

Finally, the value of a column can be determined by using a formula. A formula also enables you to combine field values or to add text to a column entry. To use a formula to determine a column display value, click the Formula radio button in the design pane. Additionally, you may want a column to display a value based on the form used. For example, examine the following formula:

@If (Form = "Questions"; "The user's question is related to" + Category; Name)

This formula will display the static text *The user's question is related to* as well as the value of the Category field if the document was created with the Questions form. If the document was created with any other form, the column will display the name of the responder, contained in the field Name.

Here is another example:

@If (Category = "1"; "How do I"; Category = "2"; "Broken"; "Not Shown")

This formula enables notes to displays an icon based on the value of the Keyword selected in the Category field. If the designer selected the Display values as icons option in the Column properties InfoBox.

The following formula shows one way to add static text to a column:

"The customer is" + CustomerName + "with" + CompanyName + "."

This formula combines the values of the CustomerName and CompanyName fields with accompanying static text, spaces, and punctuation.

If the developer wants to display the values of a keyword field in a column but used synonyms when defining the keywords, a formula similar to this one might be necessary.

@If (Software ="1"; "Lotus Notes"; Software = "2"; "MS Office"; Software = "3"; "Visio"; Software = "4"; "Netscape"; "No Software package selected")

In this example, the column would display the word (MS Office, and so on) relating to the choice the user made in the document. If the user chose Lotus Notes as the value of the keyword field when he was completing the document, for example, the column would display Lotus Notes. Without this formula, however, the column would display the number 1, since that is the value stored internally for the keyword Lotus Notes.

EXAM TIP
The preceding examples give you some idea of the types of column formulas you may see on the Application Development I exam. Remember that only one data type can be used in a single column; therefore, the @Text or other translation functions can be used to ensure all elements of a column are the same data type.

Categorize the View

Documents in a view can be organized or grouped according to a field on the form from which they were created, as discussed earlier in this chapter and in Chapter 2. This type of organization, called categorizing, can make views easier to understand and make documents easier to locate. A designer can categorize a view by creating a column that uses sorting and categorized options. The value of the column should be a field that results in an easily sortable value, such as an author or keyword. The designer can allow the user to use the categorize action to categorize a view by using a Categories field on the form being used to create the documents. This field must result in a list of words or names on which you wish to base the categories.

To create a categorized view, first make sure you have included a field on the forms that can be organized by category. Then create a column (probably far to the left) that refers to this field. Next, in the column properties InfoBox, choose Categorized on the Sorting tab. You may also want to choose *Show twistie when row is expandable* on the Basics tab. More than one column can be categorized, if necessary.

View Actions

In Chapter 3, we discussed adding Actions to forms. Actions can also be associated with views. They are created and accessed the same way. This section will describe adding actions to views. While in design mode for a view, select View . . . Action pane to display the Action pane at the top right corner of the screen. Notes provides default actions with both forms and views, which will be listed in the action pane and indicated with an asterisk (*). These default actions enable the user to perform common actions such as categorizing, moving documents to folders, and editing documents. The designer can also add custom actions to the views. To add a custom action to a view, click Create . . . Action. This will display the Action properties InfoBox, shown in Figure 4.18.

The Action must first be named; as usual, choose a unique, descriptive name for the action, as it needs to be distinguishable to be chosen from a menu. You then have the option to choose a button icon from a predefined selection. Choose whether the action will be selected from the Actions menu, button bar, or both, using the checkboxes on the InfoBox. The last option, Position, determines where the button will be displayed on the button bar, or where the Action will be listed on the Actions menu.

Some examples of useful View actions include a Create New to create a new document based on a particular form or Edit document to open and edit the selected document. To create a Create New document action, use a formula similar to this one:

```
@Command ([Compose]; "Questions")
```

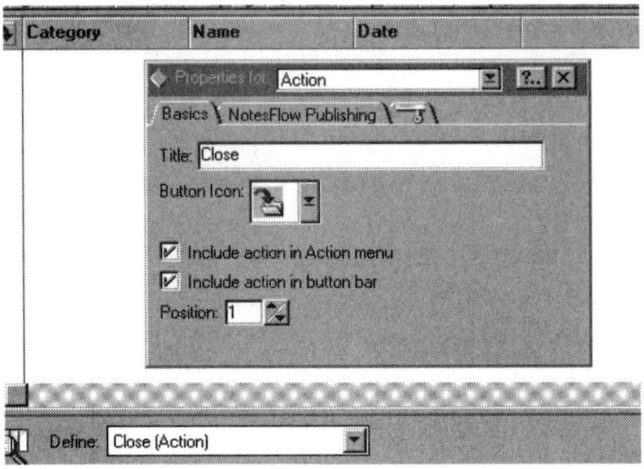

Figure 4.18 Action properties InfoBox.

This action would create a new document based on the Questions form.

Another useful View action might be to run an agent on all the selected documents, like an approval or rejection agent.

Sample Database

The sample database badly needs a view. The goal is to create a view that allows us to scan over the documents quickly to determine their status. We want to know which problems are closed and which are open. If a problem is open, we want to know who is handling the problem and when they promised to follow up on the problem. We also want to know whether someone has submitted a question but has not yet been contacted with an answer. We will categorize the view using the Categories field and then add the columns with fields and formulas. When finished creating the view, add an Action button to close the form—refer to the previous chapter if you need a reminder on Actions. Let's begin:

1. Create a new view called *First View*. Under View Properties …. Options, make sure that the Show response documents in a hierarchy checkbox is checked.
2. Create the following columns in the following order, as shown in Table 4.1:

Table 4.1 Suggested View Column Settings

Column	Contents	Special Properties
Category	@If(Categories="1";"How Do I";"Broken")	Sorting. . .Categorized Basics. . .Show twistie
No title	Status+" : responsible person: " +Support+" "+ @If(Status = "Open";"following up on: " + @Text(followup);"")	Basics. . .Show responses only
Name	@If (Form = "Q"; Name ; UName)	
Date	Probdate	Sorting. . .Click on Column Header to sort. . .Both
No title	@DocChildren("This person has not yet been contacted";"")	

Be sure to add lots of documents to your database in varying stages of completeness so that you can appreciate the power of this view. Your view should look something like the one pictured in Figure 4.19.

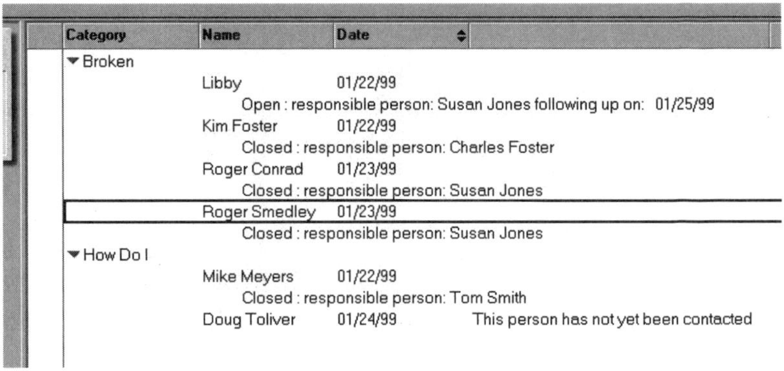

Figure 4.19 Sample view example.

Review Questions

1. Carin creates a personal view for her mail database. A few days later, she receives a laptop computer and has Lotus Notes installed. She calls the help desk complaining that her personal view has disappeared. What happened?

   a. Personal views are only cached for three days. If they are not used in this time, they are discarded.

   b. The administrator took away her rights to create personal views. When this was changed, the personal view became inaccessible to her, but the administrator can still see it.

   c. The personal view was stored in the DESKTOP.DSK on the original machine. The new laptop has a different DESKTOP.DSK, and therefore Carin cannot see the view from this machine. She can still see the personal view if she uses her old machine.

   d. The personal view was stored in the CACHE.DSK on the original machine. The new laptop has a different CACHE.DSK and therefore Carin cannot see the view from this machine. She can still see the personal view if she uses her old machine.

2. Susan creates a view for her database. When she distributes the database to her users, they all say that they cannot see the view. What did she do wrong and how can she fix it?

3. Documents are displayed in what element of a view?

4. If Ray wants to distribute a view to his users that shows only the documents that they created, what type of view should he create and what would the selection formula be?

5. Conni wants to have her users select documents to place in a container. She does not want to select these documents for her users. She does, however, want to define the columns and other formatting of the container. What type of container should she create?

   a. View

   b. Folder

6. What is a synonym for a view called? What is the difference between this and a synonym?

7. Alan wants to create a view that selects documents based on the form (JobSearch) that was used to create them. How can he choose the documents?

   a. Use the Search Builder and choose the Simple Action By Form Used.

   b. Use the Search Builder and choose the Simple Action By Form.

   c. Write a selection formula that looks like this: SELECT form = JobSearch.

   d. Write a selection formula that looks like this: SELECT form = "JobSearch".

8. Kenny created five views, but he wants the By Author view to be the first view that users see when they open the database. How can he make this happen?

   a. Choose the view name as Default on Open in the Database Properties InfoBox.

   b. Choose the Default when Database is first opened option on the View Properties InfoBox.

   c. Choose the Default when Database is first opened option in the New View dialog box.

   d. Put an asterisk at the end of the view name. The view will automatically be the default view for the database.

9. When Lynda's users open the view in her very large Company Directory database, she wants the view to open as quickly as possible. Which of the On Open options should she choose to make this happen?

   a. Go to top row.

   b. Go to bottom row.

   c. Go to last unopened document.

   d. Go to last opened document.

10. Brandon is creating a view that has the documents sorted in ascending order by date (newest at the bottom). He is choosing the Refresh options for this view. If the database is fairly small, but it changes often, which Refresh option would be best for him to choose?

    a. Display indicator.

    b. Refresh display.

    c. Refresh display from top row.

    d. Refresh display from bottom row.

11. Missy created a view with a solid colored background, but her users complain that it is hard to distinguish one document from the next. What can she do to make it easier to distinguish the documents from each other?

    a. Use a different color for alternate rows. This option is available in the Column Properties InfoBox.

    b. Use a different color for alternate rows. This option is available in the View Properties InfoBox.

    c. Use 1 1/2 line spacing between rows. This option is available in the Column Properties InfoBox.

    d. Use 1 1/2 line spacing between rows. This option is available in the View Properties InfoBox.

12. Basil is changing the look of a view created by another designer. He notices that the column headings blend right in with the documents. The other views in the database have gray, box-like headings with a 3-D effect. How can Basil make this view look more like all the others?

    a. He can't. The view headings can't be changed after the view has been saved.

    b. He can change the Heading Style for each column in the Column properties InfoBox.

    c. He can choose Simple Headings in the View Properties InfoBox.

    d. He can choose Beveled Headings in the View Properties InfoBox.

13. Paula is planning to display the Customer Address information in the By Customer view of her database. The Customer Address information will be between three and five lines. She wants to make sure that all the information shows up, but she also wants to make sure that users don't have to see blank lines. What options will she enable to do this?

14. Millie creates a view and types in the following column headings: Company Name, Company Address, Product Order, and Order Status. She decides that it would look better to have each title appear on two lines, but when she puts a return at the end of the words, she cannot read the second word in the Column headers. What option can she enable or change to make the Column titles fit better?

    a. Change the Lines per Heading option to 2 in the Column Properties InfoBox.
    b. Change the Lines per Heading option to 2 in the View Properties InfoBox.
    c. Enable the Multi-Line option for Headings in the Column Properties InfoBox.
    d. Choose Fit to Content in the Column Properties InfoBox.

15. Nicole told the people that were developing her application that she wanted to be able to change the width of the columns to view her data more easily. They told her that this was not an option. Is this true or false? If it is false, what could they do to enable this option?

16. Greg choose the option Click on Column Header to sort in the Column Properties InfoBox. Which of the following are the options that he can choose with this?

    a. Ascending
    b. Descending
    c. Both
    d. Change to View

17. Rick wants to create a column that will display this information: The Company *CompanyName* has ordered *Quantity* of the *WhichBook* book. How can Rick make this information display in the Order Information column?

18. In another column, Cindy wants to display the status of the document—Approved, Pending, or Rejected. If the Status field is a keyword field that does not use synonyms, what formula would display the correct value?

19. In the example above, how would Cindy have to change the formula if the keyword field Status uses the following synonyms: Approved = 1, Rejected = 2, Pending = 3?

20. Still using the example above, could Cindy show an descriptive icon, such as a smiley face for approved, a sad face for rejected, and a neutral face for pending? If so, how?

21. Jordan wants to categorize the documents in the By Author view by the Author of the document. How can she do this?

22. Jordan categorized the documents in the By Author view, above, but when the categories are collapsed, she has to click the Expand SmartIcon to expand them. She wants to know if there is another way to do this.

    a. No. The Expand SmartIcon is the best way to expand the documents in the view.
    b. Yes. She can choose Expand Documents Automatically from the View Properties InfoBox.
    c. Yes. She can choose Show Twistie when Row is Expandable from the View Properties InfoBox.
    d. Yes. She can choose Show Twistie when Row is Expandable from the Column Properties InfoBox.

23. Hannah opens a document from the new view that was created for her database. The document looks different than when she created it. All the static text and graphics on the document are different. What might have happened? She only uses one database with two forms that she uses to input data.

    a. The developer used a form formula in the view to choose a different form to display documents.

    b. The developer forgot to choose to store the form in the document to show the correct form.

    c. The form was deleted from the database and the default form for the database was used instead.

    d. The designer used hide-when formulas on the form to make it appear different when opened for reading.

24. GJ wants to create a view that will display all documents created with the form Patient Profile and all the responses to these documents, at any level. What formula would he use for the selection formula?

    a. SELECT form = "Patient Profile" & @AllDescendants

    b. SELECT form = "Patient Profile" | @AllDescendants

    c. SELECT form = "Patient Profile" | @IsResponseDoc

    d. SELECT form = "Patient Profile" & @AllChildren

Answers

1. Answer c is correct. Personal views are stored in the DESKTOP. DSK file. This file is stored in the \NOTES\DATA directory on the user's machine. The personal view can be stored in the server copy of the database if the Create Personal Folders/Views right is checked on the ACL dialog box.
2. Susan probably created a Private view by forgetting to click the Shared option on the Create View dialog box. To fix this problem, Susan must create a new shared view and copy the columns and selection formulas from the original view. She can then delete the original view.
3. Documents are displayed in rows.
4. He should create a Shared, Personal on First Use view. The Selection formula might be something like this: SELECT @Contains (@Author; @UserName).
5. b. She would create a folder. The designer chooses documents for a view; the user chooses documents for a view.
6. A synonym for a view is called an alias. There is no functional difference between an alias and a synonym.
7. Both answers a and d are correct. Either option has the same result: Use the Search Builder and choose the Simple Action By Form Used OR Write a selection formula that looks like this: SELECT form = "JobSearch".
8. b. Choose the Default when Database is first opened option on the View Properties InfoBox.
9. d. Go to last unopened document is the best answer.
10. d. Refresh display from bottom row is the best answer.
11. Both b and d are available. Use a different color for alternate rows and use 1 1/2 line spacing between rows. Both of these options are available in the View Properties InfoBox.
12. d. He can choose Beveled Headings in the View Properties InfoBox.
13. She can choose to allow five lines per row and then enable the Shrink rows to content. This makes the default row height five rows, but if some of the documents do not use all the rows, the row will only be the number of lines necessary. These options are available on the View Properties InfoBox, Style tab.

14. b. Change the Lines per Heading option to 2 in the View Properties InfoBox.
15. False. To enable your users to change the width of columns at will, choose the Resizable option on the Basics tab of the appropriate Column Properties InfoBox.
16. a, b, c, and d—all answers are correct.
17. The formula for the Order Information column would be "The Company" + CompanyName + "has ordered" + @Text (Quantity) + "of the" + BookName + "book."
18. @If (Status = "Approved"; "The proposal has been Approved"; Status = "Rejected"; " The proposal has been Rejected"; "The proposal is Pending").
19. @If (Status = "1"; " The proposal has been Approved"; Status = "2"; "The proposal has been Rejected"; "The proposal is Pending").
20. Yes. She would have to choose the option to Display values as icons on the Column Properties InfoBox for the column and write the following formula: @If (Status = "Approved"; 85; Status = "Rejected"; 86; 87). The values for the icons available for columns are available in the Notes Help database or in the *Application Developer's Guide*.
21. Use the Simple Function Author(s) Simple Name to define the column value. In the Column Properties InfoBox, choose the Sort tab. Choose a Categorized type instead of standard and sort Ascending.
22. d. Yes. She can choose Show twistie when row is expandable from the Column Properties InfoBox.
23. Actually, a, c, or d could be true. A designer probably used a form formula to choose the form to display the documents in the view. This is done on the View Properties InfoBox. The designer also might have deleted the original form from the database, either intentionally or accidentally. If the form used to create the document is not available, the default form for the database is used. D would only be true if the form changed back to the way it looked previously when she tried to edit the document.
24. b. SELECT form = "Patient Profile" | @AllDescendants.

CHAPTER 5

Complete the Application

The Application Development I exam expects the developer to be able to put the finishing touches on an application. These include some of the rudimentary security settings in the Access Control List, the About this Database and the Using this Database documents, and the Navigators that make the database more user-friendly. These finishing touches to the database are discussed in this chapter.

Chapter Objectives

You should be able to answer questions based on the following objectives after reading this chapter:

- Limit access to the database using the Access Control List.
- Write an About this Database document.
- Write a Using this Database document.
- Create a graphical Navigator in the database.

Introduction to Database Security

Although Security is not a major focus of the Application Development I exam, the exam does touch on the topic. Notes offers many security levels for an application from field level security, to User ID authentication with the server, to encryption. One of the main security components in databases is the *Access Control List* (ACL). The Application Development I exam deals with the ACL for the topic of security. This section describes the functions the ACL performs for securing the database and the levels of ACL.

The ACL settings provide a variety of functions for the database. First, they determine who can access the database. It is good practice to set the Default access to the database to No Access so that users have to be either explicitly listed or in an explicitly listed group to gain access to the database. Second, the ACL determines which functions the user can perform. For most users, this includes reading, writing, and editing documents. Finally, the ACL assigns the roles and other specific job tasks of the users, which in turn can be used to determine the views and forms the users can access and use.

To view the ACL for a database, right-click the database icon and choose Access Control . . . or select File . . . Database . . . Access Control from the menu. This displays the Basics panel of the ACL dialog box, shown in Figure 5.1.

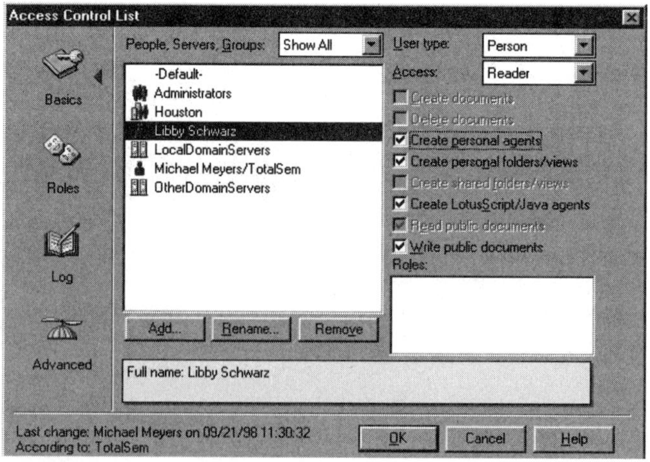

Figure 5.1 ACL dialog box.

The Basics panel of the ACL dialog box enables the manager of the database to change, add, or delete the access and roles of the people, servers, and groups that need to access the database. To add a person, server, or group to the ACL, click the Add . . . button at the bottom of the Basics panel. A prompt is displayed, which enables you to type in a name or select it from the Public *Name and Address Book* (NAB). When given the choice, it is best to select the name from the NAB to prevent mistakes. After adding the person, server, or group, select the level of access and the user type from the drop-down lists on the right side of the dialog box. The levels of access for a database are described in Table 5.1.

NOTE

When you create a new database, you are automatically the manager of the database until you change these rights. As the developer, you set the ACL first. This activity will be an ongoing part of the database manager's job, however, so you will probably change these rights. In addition, the Default access to the database is set to designer when the database is created. This access is in effect for anyone not specifically listed in the ACL.

Table 5.1 Security access levels

Access Level	Capabilities
Manager	There should always be at least one manager of the database, who has the ability to change the ACL settings. In addition, the manager can change replication settings, locally encrypt the database, and delete the database. Manager access also includes all the abilities of the other access levels.
Designer	Database designers have the ability to create, change, and delete all design elements of the database including forms, fields, views, and actions. In addition, the designer can create a full text index for the database and has all the abilities of the lower access levels. Remember that the default access to the database is set to designer and must be changed before testing the database.

continues

Table 5.1 Continued

Access Level	Capabilities
Editor	Users with Editor access to the database can read, create, and edit all documents in the database. Editor access should be limited to one or two users, if possible, to limit the possibility of Save conflicts caused by multiple users editing the same documents at the same time.
Author	Author access to the database gives the ability to read and create documents. Additionally, the user has the right to edit any documents that contain their name in an Authors field. Usually, this includes any documents that the user created, but the designer can also create formulas to add other names to the Authors field.
Reader	Readers of the database can read documents in the database but cannot create or edit documents. This access is useful for documentation or help databases.
Depositor	A depositor can create new documents in the database but cannot read or edit documents. This can be useful for surveys, ballots, or evaluations.
No Access	Users with No Access to the database cannot access the database. In addition, they cannot add the database icon to their workspace.

EXAM TIP

This description of the ACL is limited to the questions on the Application Development I exam. For more discussion of database security, refer to Chapter 11, which describes more security aspects of the database, as tested on the Application Development II exam.

Database Help

A database developer must also provide help documentation to the users of the database. The two documents provided by Notes to serve this purpose are the About this Database and Using this Database documents. A user must have designer or manager access to edit these documents. All users can read these two documents by selecting them through the Help menu. The About document can also be accessed by clicking the About . . . button on the Open Database dialog box. This allows the user to see a description of the database and its purpose before opening it or adding the icon to the workspace.

To create the database help documentation, choose View . . . Design to access the design elements. Then select Other from the design elements listed in the navigation pane. The About and Using documents should be available in the View pane. When opened, each of these documents is a blank slate that is essentially a rich text field. This rich text field can contain formatted text, graphics, hotspots, attachments, doclinks, buttons, and other objects.

The About this Database document is usually a brief description of the purpose and audience of the database. Other information that might be added to this document includes:

- Contact information for the database manager
- Guidelines for the use of the database
- Major functions or features of the database
- Date the database was created and implemented

The About document can be accessed by clicking Help . . . About this Database. The About document can also be launched automatically, as we discussed in Chapter 1. A designer can also create an action button to access the About and Using documents, as this is information users may need in order to use the database successfully. The users will receive an error if they click on the About this Database document and no About document exists in the database. For this reason, it is important to create an About document for every database you develop.

The Using this Database document is usually the main help that the users receive for the database. You may also have added other help such as field help, field pop-up help, and doclinks to the help. This document, however, provides an easy way to define the use and organization of each form, view, agent, and action. In addition, you may want to describe the work process that the database was designed to perform. Again, you can include links to other documents, graphics, attachments, formatted text, buttons, and other objects to make the help documentation as complete and thorough as possible. Users will access the document by clicking Help . . . Using this Database. As mentioned previously, you may also want to create an action button or other link to make this document easily accessible to the users.

Database Navigators

A Navigator, which is a new feature to Notes R4.X, is a graphical table of contents for a database. Figure 5.2 shows an example of a graphical Navigator that allows users to choose views, forms, and actions. It allows users to open views, forms, and documents without having to use the menus or lists of views and forms. A Navigator uses links, similar to doclinks, to move the user to various areas of the application. The Navigator can be displayed in the Navigator pane on the right side of the screen, or it can be displayed in its own window. The default Navigator, with which you are probably familiar at this point, has magnifying glass icons to indicate views and folder icons to indicate folders. The graphical Navigators that you can create can make finding views, forms, and documents in the database more intuitive. Figure 5.2 shows the interesting Navigator from the Lotus Help database.

Figure 5.2 Navigator example.

To create a Navigator, choose Create . . . Design . . . Navigator from the menus. This opens an empty design pane and blank Navigator. Right-click in the upper pane to choose Navigator properties. The Navigator properties InfoBox, shown in Figure 5.3, enables you to give the Navigator a name, a background color, and a default initial view.

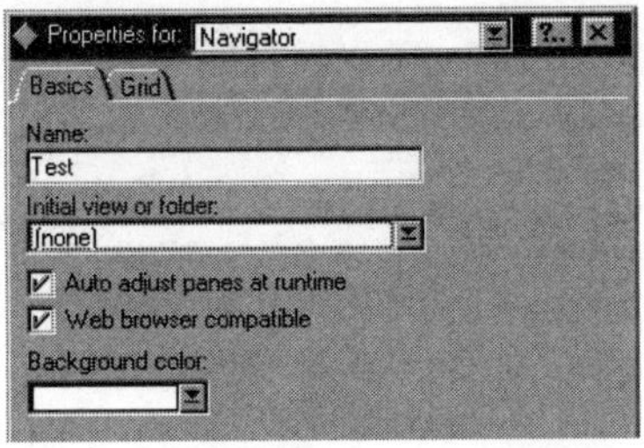

Figure 5.3 Navigator properties InfoBox.

Give the Navigator a unique name, using any characters you like. As usual, only the first 32 characters appear on menus. The Initial View or Folder option allows a default view or folder to be displayed when the Navigator is displayed. This is especially useful if the Navigator is being automatically launched, using options on the Database Properties InfoBox. To resize the panes automatically to accommodate the graphical items in the Navigator, enable the Auto Adjust panes at runtime option.

You can add a colored background to the Navigator using the drop-down list on this InfoBox. Additionally, you can add other graphic elements such as graphic buttons and hotspots to the Navigator. You also have the ability to copy graphics from other programs and either paste them into the Navigator using Paste or choose Create . . . Graphic Button from the menus. To create a hot-spot, select Create . . . Hotspot Rectangle from the menus. To add an action to the button or hotspot, use the design pane to add a simple function, @Function formula, @Command formula, or LotusScript formula.

To display the Navigator, either choose View . . . Show . . . and the name of the Navigator, or use the Database Properties InfoBox to launch the Navigator automatically. You can also launch the Navigator using an Action in a view or form. An example of an action to open a Navigator would be:

```
@Command ([OpenNavigator]; "MainNavigator"; 1)
```

This command opens the Navigator named Main Navigator in its own window. The parameter *1* indicates opening the Navigator in its own window; a parameter of *0* would open the Navigator in the left-hand side Navigator pane.

Sample Database

Although we cannot do anything with security at this point, we can go ahead and create some groups that we can use later. Select File . . . Database . . . Access Control to edit the ACL. Add two people groups called *Training* and *Support*. Keep their defaults with the exception of turning off the Create Personal Agents checkbox. Be aware that adding groups to the ACL does nothing if the same groups do not exist in the Notes Name and Address book. We are adding groups here for example purposes only.

Let's make a Navigator for the sample database. This Navigator will allow a person to create a question, create an answer, or exit to the Lotus Notes Workspace. Select Create . . . Navigator and name the new Navigator *First*. Create three hotspot buttons, called Question, Answer, and Exit. Give them the following formulas:

Question: `@Command([Compose]; "Questions")`

Answer: `@Command([Compose]; "Answer")`

Exit: `@Command([FileCloseWindow])`

To make the new Navigator start automatically when the database is opened, go to the Database Properties and select the Launch tab. Change the Launch properties as shown in Figure 5.4.

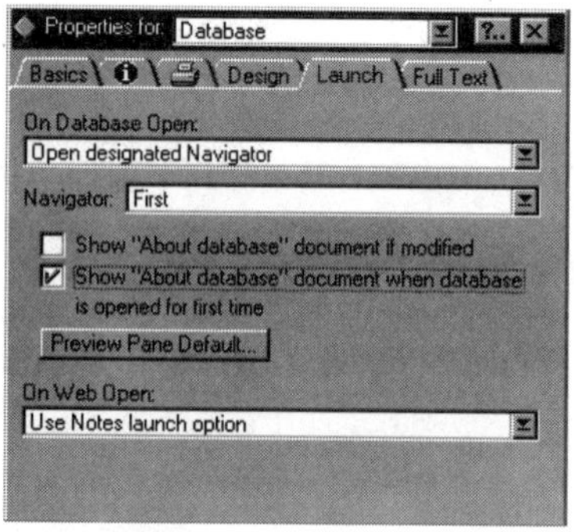

Figure 5.4 Sample database Launch InfoBox.

Save and close the database and re-open it to see the new Navigator in place (see Figure 5.5).

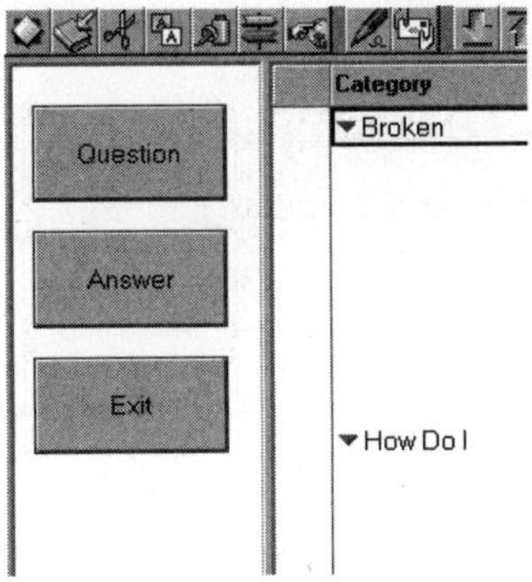

Figure 5.5 New Navigator in place.

Review Questions

1. Juliana puts the new database she has designed out on the server without changing the default ACL. When she looks at it a few days later, she notices that someone has changed the forms and views she has created. Why did this happen?

   a. The Everyone group has Manager access by default.

   b. The Everyone group has Designer access by default.

   c. The Default group has Manager access by default.

   d. The Default group has Designer access by default.

2. Mike gives the Doctors group in his Patient Profile databases on ServerA and ServerB Editor access to the database. After a few days, he notices that there are a lot of Replication and Save conflicts in the database. Is there anything he can do to prevent these errors?

   a. Because the database is on two servers, there is nothing he can do.

   b. He can enable Merge Replication Conflicts in the Form Properties InfoBox.

   c. He can give the Doctors group Author access instead.

   d. He can disable the database on one of the servers and enable Merge Replication conflicts on the Database Properties InfoBox.

3. You are talking with Laura, another application developer in your team. She complains that she gets five to 10 calls every day from users asking for help in using her database. You give her the following advice:

   a. Tough luck. We all have to deal with those calls. It's part of your job.

   b. You could create pop-up help for the fields in the database by making some of your static text hotspot pop-up.

   c. You could enable field help in the Field Properties InfoBox.

   d. You could create help for the database by editing the Using this Database document.

4. Denise wants to create About this Database and Using this Database documents. She has looked on all the menus and says that she still cannot find the right menu to choice. Can you help her?

   a. They are under Design . . . Help Documents.
   b. They are under Create . . . Design . . . Help Documents.
   c. They are already created. If you look in the Design . . . Help documents view, you can open and edit them.
   d. They are already created. If you look in the Design . . . Other view, you can open and edit them.

5. An About this Database document or a Using this Database document *cannot* contain which of these elements?

   a. Tables
   b. DocLinks
   c. Attachments
   d. Graphics hotspots or buttons

6. Heidi is creating a complex database that allows users to schedule train trips through her travel agency. She wants to help users choose the correct destinations by using graphics. What can she use that combines graphics with navigating to the correct forms and documents?

7. Charlie wants users to see the graphical Navigator that he created as soon as they open the database. What does he need to do to accomplish this?

   a. On the Navigator Properties InfoBox, choose the Initial View or Folder Option.
   b. On the Navigator Properties InfoBox, choose the Navigator in the On Database Open option.
   c. On the Database Properties InfoBox, choose the Initial View or Folder Option.
   d. On the Database Properties InfoBox, choose the Navigator in the On Database Open option.

Answers

1. d. The Default group has Designer access by default.

2. Both b and d are correct. He can enable Merge Replication Conflicts on the Form Properties InfoBox. This only helps him in situations where users are on different replicas of the database and changing different fields on the documents. He can also choose to give the Doctors group Author access so that they are only editing the documents they created instead of editing all documents in the database.

3. b, c, and d are all possible answers. You could create pop-up help for the fields in the database by making some of your static text hotspot pop-up, or you could enable field help in the Field Properties InfoBox, or you could create help for the database by editing the Using this Database document.

4. d. They are already created. If you look in the Design . . . Other view, you can open and edit them.

5. None of the answers are correct. An About this Database or Using this Database document uses a Rich Text field, and all of the elements shown in the question can be used on these documents.

6. She can create a graphical Navigator to help her users navigate through the database.

7. d. On the Database Properties InfoBox, choose the Navigator in the On Database Open option.

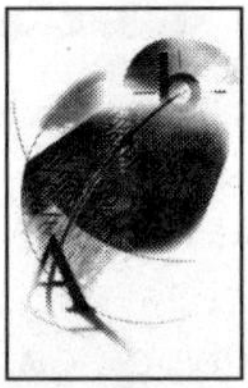

CHAPTER 6

Advanced Applications and Workflow Database Design

In this chapter, we begin discussing the topics covered by the Application Development II exam. The topics listed in the official Lotus-published Exam Guide may be confusing to many testers, as they appear to be the same topics covered in the Application Development I exam. Essentially, the Application Development II exam covers the same major topics of forms, formulas, actions, views, and security. They are covered in deeper detail, however, with more attention to advanced functionality. Some additional topics are also addressed, such as workflow and security. The next six chapters will explain these topics in detail.

The topics covered in these chapters assume that you have read the first section of the book and have experience developing Notes databases and applications. In each chapter, there will be a limited review of the topics covered on the Application Development I exam. The Application Development II exam assumes that the tester already has this level of knowledge and can build on it for the advanced topics. The chapters follow this assumption by reviewing and then building on the Application Development I knowledge of each topic to move to the more advanced functions required by Application Development II.

In Application Development I, we were mostly concerned with developing single, stand-alone databases. The type of database that we concentrated on in Application Development I was usually a discussion database. Within the Application Development I chapters, we also defined each of the other types of databases including broadcast, tracking, reference, and workflow. Reference, tracking, and discussion databases could be developed using the functions and skills learned and tested in Application Development I. In Application Development II, however, we will build onto these database types by focusing on the workflow type of application and the capability of Lotus Notes to incorporate mail functionality into the databases. The workflow applications are usually multi-database applications that include both a workflow database and discussion, reference, tracking, or broadcast database. Some examples of workflow applications include scheduling conference rooms or other resources and automating purchasing or other sales tasks. In this chapter, therefore, we will begin to examine the planning of any advanced Notes application and the developing of a workflow application, including workflow terminology, Notes mail options, and structure.

Chapter Objectives

You should be able to answer questions based on the following objectives after reading this chapter:

- Plan advanced applications
- Understand workflow terminology
- Develop workflow applications
- Mail-enable Notes forms
- Use the @MailSend function
- Use Notes Reserved fields for mailing

Plan Advanced Applications

When a designer receives a user request to develop an application, the designer usually begins by planning what will be developed and by discussing the application design and function with the user. This planning stage serves two purposes. First, it allows the user requesting the application to reassess the business process that is being automated. Second, it ensures that the designer and the user community both agree on the required forms, views, and functions used to automate the process. In this way, you avoid designing an application that does not meet the needs of your user community and will need to be redesigned to be useful. The planning stage is important for all databases that you design. It is even more important, however, to work in this method when you are designing workflow applications or other multi-database and advanced feature applications. The steps for designing the architecture of an application and developing it are outlined here.

1. Discuss the current business processes and structure with the potential users to determine the needs of the organization. This step includes interviewing the users, researching the current process that uses paper or computer-based procedures, and determining the number and types of databases needed.
2. Plan the database and/or application on paper, including forms, fields, views, and functions for each of the databases. Ensure that the flow of the business process (workflow) is mapped out clearly, including how this is implemented in the application structure. A good way to map out this portion of the database is to use a process flow diagram.
3. Review the design with the potential users. This gives the users the opportunity to make design changes and requests. Implement the changes in the paper plan of the application.
4. Create the prototype .NSF file on your workstation. Add the forms, fields, views, and functionality to the application.
5. Test the prototype with a small group of test users in a separate testing environment, if one is available.
6. Implement any changes and use the updated prototype as a template for the final version of the database.
7. Move the final version of the database to the server after implementing an appropriate Access Control List and other security measures agreed to by the database manager, designer, and the system administrator. In addition, when introducing the application to the core group of users, provide training and documentation, starting with About and Using documents.

These steps are intended to ensure that the application development process is completed within the timeframe allotted and without requiring the developer to redo aspects of the application that the users do not find acceptable. The reason for developing applications is to make the business process flow more smoothly. The application developer must always keep this in mind when completing an application. The Application Development II exam requires the developer to know and understand these planning steps.

The planning of an application is especially important when it is a workflow type of application, because these types of applications require the most knowledge of the business processes on the part of the developer.

Creating a Workflow Application

When creating a workflow application, the designer first needs to understand the business process that he or she is automating. A *business process* is defined by Lotus as the steps used to reach a business goal. The activities and interactions required by the business process can then be automated by Notes using workflow functions. *Workflow* can be defined by the activities (work) and interactions (flow) required to complete that business process.

The workflow application can be designed along a set of basic models. The three main approaches are the sending model, the sharing model, and the hybrid model. When using the sending model, the process uses actions to route forms or data from the database to the user. The form or information arrives in the user's mail file, after which the user can act on it. This model puts some of the responsibility for the dissemination of information on the developer. The best use of this model is a process that is not continuous or expected, also called an *ad hoc* process.

When using a sharing model, however, the information is stored in a database. The users have to open the database to complete the forms or read data. The sharing model puts the responsibility on the user to go to the database to complete the business processes. This process is best used when habit reminds the users to go to the database. Continuously running processes, for example, are best suited to this model.

The hybrid model combines the best of both of these ideas. The information or form remains in the database and a reminder or link to the data is sent to the user. In this way, the user is given the information they need without the wasted effort of checking the database. Furthermore, the designer has the ability to save space on the server and maintain forms only in the database.

The sharing model uses the functionality described in the Application Development I portion of the text. The sending and hybrid models require the ability to mail information to the user. This is the basis of the workflow application—automating the flow of the business process by sending the information that needs to be acted upon directly to the user.

When determining how to create workflow applications, the tools within Notes that you can use include actions, agents, and the mailing functions of Notes. The latter is enabled using special @Functions and reserved fields. Together these elements can be combined to automate the tasks in the business process. Alone, these elements do not completely automate tasks.

Automate Mailing Within Notes Applications

The main function that is automated in workflow applications is the routing of forms and information using the mail functions of Notes. Mail is usually routed based on some kind of trigger, either manual or scheduled. In the case of a manual trigger, a user might click a button that contains an action or perhaps enter a particular value in a field and then save it. In the case of a scheduled or automated trigger, the document could be routed based on certain conditions (whether a document has been saved or edited) based on the value of a field or based on the day or time. In the question tracking database that we created earlier, for example, the user might click a Submit Question button that routes the question they have asked to a specific person based on the topic of the question. Similarly, the questions might route automatically every hour to the technician on call at that time.

When automating mail routing, the mail can be routed either to a user's mail file or to a mail-in database. With any mail routing, three main requirements must be fulfilled for the routing to occur correctly:

- The form that is being used to route must be mail-enabled. This can be controlled either by the designer (using a formula) or by the user.
- The message must have a proper address included. Addressing is usually accomplished by a SendTo field or by the parameters of an @MailSend formula, both of which are described later in this chapter.
- A route to a mail-enabled destination must exist. When the destination is a user's mail file, the route is defined by that user's

unique person document in the *Public Name and Address Book* (NAB). When the destination is a mail-in database, the route is defined by a mail-in database document created in the NAB.

Mail Enabling a Notes Form

You can enable a form for mailing in one of three ways. You can set the mail options in the Form Properties InfoBox, you can add a MailOptions reserved field to the form, or you can use an @MailSend function to send the document.

When designing a form that will be mail-enabled, take into consideration how the document will be viewed in the target database, whether that is the user's mail file or a mail-in database. To display the document using the form that was used to create it, the form must be available in the target database. Two options to make the form available in the target database include storing the form in the document or copying the form to the target database. To store the form with the document, enable the *Store form in document* option in the Form Properties InfoBox, shown in Figure 6.1.

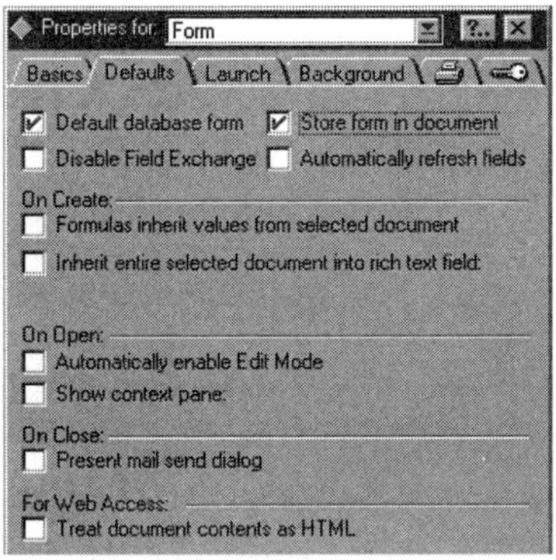

Figure 6.1 Store form in document option.

Setting the Mail Options

To set the mail options for the form, open the Form Properties InfoBox and click the Defaults tab. The option that you will select is On Close: Present mail send dialog, as shown in Figure 6.2.

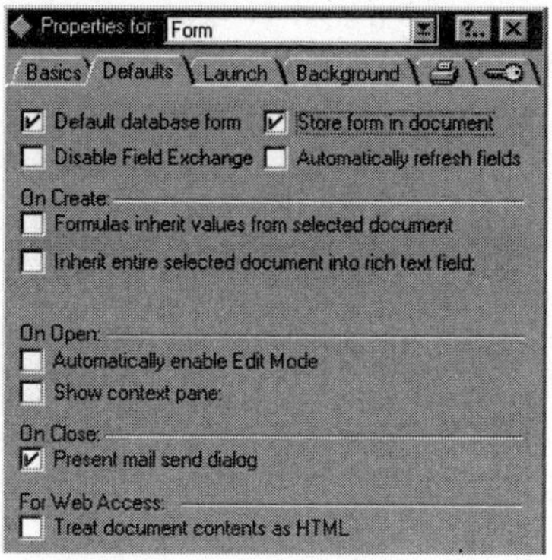

Figure 6.2 Form Properties InfoBox showing the mail option.

When this option is enabled, the user will be given the ability to mail, save, sign, encrypt, or cancel the document at close time. Figure 6.3, shows an example of the Mail Send dialog box that is presented to the user. The Mail Send dialog box is the best option to use for mail-enabling the form if sending the document is optional.

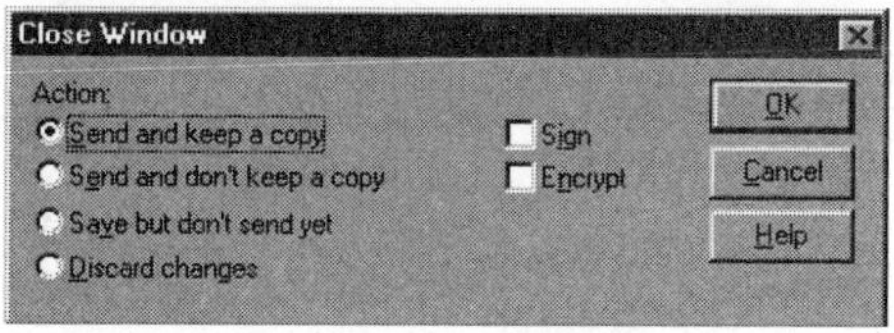

Figure 6.3 Mail Send dialog box.

Using the MailOptions Reserved Field

Another way of mail-enabling a Notes form is to use a MailOptions reserved field. A *reserved field* is a field whose name is predefined within Notes and can only be used to perform a specific function. The MailOptions reserved field specifies whether documents should be mailed automatically. The MailOptions field is usually a hidden field on the form with a computed value that resolves to a text string of either *1* or *0*. If the value is *1*, the document will be mailed automatically at close time. If the value is *0*, the document will not be mailed.

NOTE

If the MailOptions field is created and evaluated in the form, and the mail options in the Form Properties InfoBox are set, as discussed, the MailOptions field value will override the properties.

Even when the Mail Send dialog box is displayed to the user, the value of the MailOptions field takes precedence, regardless of the choices the user makes in this dialog box.

To route a document, the form must specify recipients. With either of the two options discussed previously, the additional reserved fields necessary for addressing mail must be included on the form. The fields that can be used for mailing include the SendTo, CopyTo, and BlindCopyTo fields. These fields can contain multiple text values. The text values can be obtained using fields of the following data types: text, keyword, author names, reader names, or names. If these reserved mail fields are not included on a mail-enabled form, an error dialog box displays and the form is not mailed (see Figure 6.4).

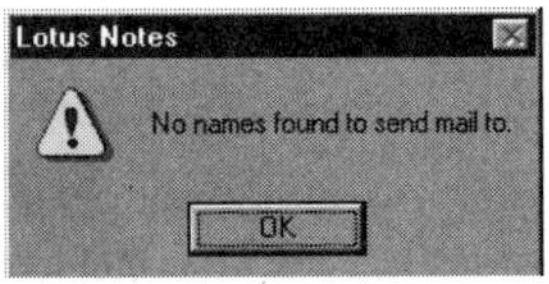

Figure 6.4 Mail Error dialog box.

Using the @MailSend Function

The third option for mail-enabling a form is adding an @MailSend function to the form. The @MailSend function can be used to route mail automatically based on the parameters included in the function. The @MailSend function can be included in most design elements such as field formulas, actions, agents, and buttons. It cannot, however, be included in elements such as the Window Title formulas, column formulas, hide-when formulas, and view selection formulas. The syntax for the @MailSend function is:

@MailSend (SendTo; CopyTo; BlindCopyTo; subject; remark; BodyFields; [flags])

If the @MailSend function is used without the parameters, which are optional, the open or selected document is sent using the reserved SendTo, CopyTo, and BlindCopyTo fields described previously. If no parameters are used with the @MailSend function, the form must have a SendTo field or an error is generated.

If the @MailSend function is used with the parameters, a new mail message is compiled and sent. The new mail message will use the values of the parameters, which are described in Table 6.1.

Table 6.1 @Mailsend Parameters

Parameter	Description
SendTo	The message will be sent to the users listed in this parameter. The parameter value must result in text or a text list.
CopyTo (optional)	The message will be copied to the users listed in this parameter. The parameter value must result in text or a text list.
BlindCopyTo (optional)	The message will be blind copied to the users listed in this parameter. The parameter value must result in text or a text list.
Subject	This parameter completes the subject field of the memo.
Remark (optional)	This parameter adds text to the beginning of the body field of the memo.
BodyFields (optional)	This parameter defines the fields in the current document that will be sent in the mail memo. The fields listed in this parameter can be of any data type. To add multiple fields, use the Notes list concatenation format of list items, enclosed in quotation marks and separated by colons, such as "date" : "user" : "software". When you include multiple fields, they are listed on the mail memo in the order stipulated. If you do not stipulate any fields in this parameter, use the IncludeDocLink flag to attach a doclink to the current document.
[flags] (optional)	The optional flags are described later in this chapter.

When you use the parameters of @MailSend, be sure to leave null strings for any values that you do not define, but that come before values that you do define. When you want to send a form containing the fields user, description, and software, for example, but you do not have any users that need to be copied or blind copied, you have to use null strings. The formula here shows the use of the @MailSend function in this situation, assuming you want the words *New Software* as the Subject, and *Please order this software for* the listed user as the Remark:

```
@MailSend ("Libby Schwarz"; "" ; "" ; "New Software"; "Please
order this software for: " + User; "software")
```

The null strings in this case hold the place of the optional parameters. If only the SendTo parameter is being used, you can leave out the null strings. In most cases, however, you will choose to send a specific field, which requires all the previous parameters to be completed with either a value or with a null string.

In addition to the parameters described here, optional flags are allowed when using the @MailSend function. These flags correspond to some of the mail delivery options available with new mail memos, such as signing, encryption, priority, return receipt, and delivery report. The flags allowed in the final parameter are described in Table 6.2. Note that all the flags are considered optional parameters and are not necessary for the @MailSend function to be evaluated correctly. In addition, you only need to include the flags you need; null values are not required within the flags.

Table 6.2 @Mailsend Flags

Flag	Description
[Sign]	This flag appends the user's electronically produced signature to the mail memo. Signing is described in Chapter 11.
[Encrypt]	This flag encrypts the document using the recipient's public key, found in the NAB. The recipient's private key, found in the user's ID file, must be available to decrypt and read the document. Encryption is described more fully in Chapter 11.

Flag	Description
[PriorityHigh], [PriorityNormal], or [PriorityLow]	This flag defines the priority of the mail message. Without a priority flag, Notes uses the default normal priority to send the message. The priority flag determines how the mail message is sent when the recipient's mail file is on a different server. This flag determines how quickly the message is routed among servers.
[ReturnReceipt]	This flag forces Notes to send a message to the sender when the recipient opens the document.
[DeliveryReportConfirmed]	This flag forces Notes to notify the sender when the document is received in the recipient's mail file.
[IncludeDocLink]	This flag includes a doclink in the mail memo that points to the document open or selected at the time the @MailSend function was run. Use this flag if you choose not to specify any fields in the BodyFields parameter.

The formula here builds on the previous formula, but includes some of the optional flags:

```
@MailSend ("Libby Schwarz"; "Brian Schwarz" ; "" ; "New Software"; "Please order this software for: " + User; "software";[PriorityHigh]:[ReturnReceipt])
```

This formula sends a newly composed mail message to Libby Schwarz with a copy to Brian Schwarz. The subject of the memo is New Software with the body reflecting the remark and the software field. The message is sent with high priority and a return receipt is sent to the user that created the document when Libby and Brian open the message.

Note that the flags are one parameter of the @MailSend function. When more than one flag is used in the parameter, it should use a list format, which requires that the flags be concatenated using colons.

TROUBLESHOOTING TIP

If the database is not on a server and cannot find a path to a server, the mail functions will not work. Notes will display an error and the user will be unable to send the document. If sending is required by a Mail options = 1 field, the user will be unable to save the document as well.

Creating a Mail-In Database

To send mail messages to a database other than a user's mail file, you must first create a mail-in database document for the database in the *Public Name and Address Book* (NAB). This document serves to define the route to the database. To create a mail-in database document, open the NAB to the Servers . . . Mail-In Databases view. Click the Add Mail-In Database action button to display the Mail-In Database form, shown in Figure 6.5.

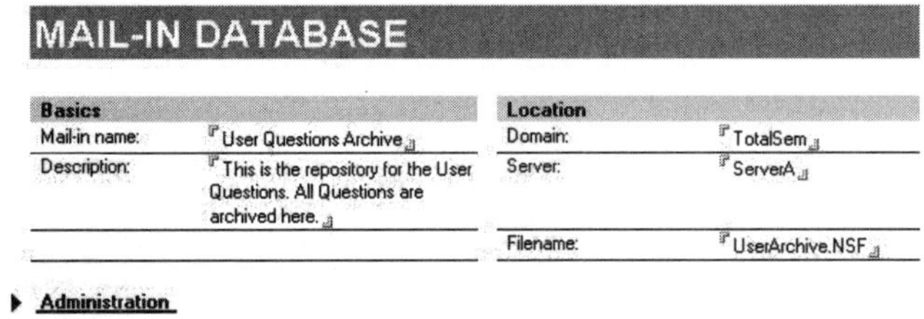

Figure 6.5 Mail-In Database document.

Type in the name used to address documents to the database in the Mail-in name field. Add a description of the database purpose in the Description field. Then provide the full Location information for the database in the domain, server, and file fields. Without these fields, the mail will not be able to find a route to the database.

NOTE

Most often, the System Administrator will have to create the Mail-In Database documents for you. In most environments, other users, including developers, do not have the necessary rights to create these documents in the NAB.

Once the Mail-In Database is created, the database can be seen in the Mail-In Databases view in the NAB as shown in Figure 6.6.

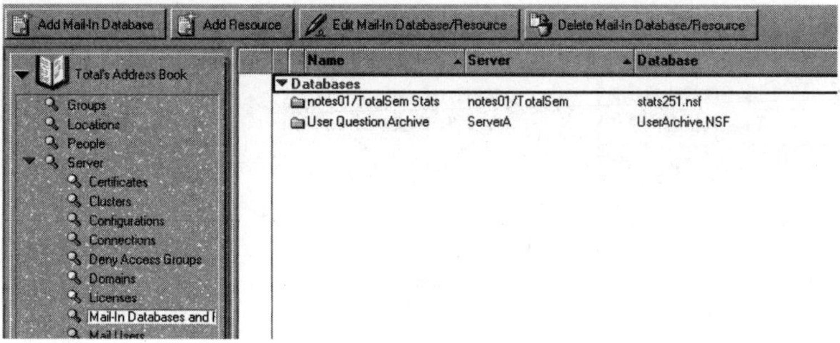

Figure 6.6 Mail-In Database view in the NAB.

When using the @MailSend or other mail-enabling tool, the recipient receives an email message that looks like Figure 6.7. When the user clicks the DocLink, Notes opens the document at the recipient's workspace. Using the options described, you can create a workflow application that mails documents either to individual users or to mail-in databases. The documents can be mailed using form-enabled mail options, the MailOptions reserved field, or the @MailSend function.

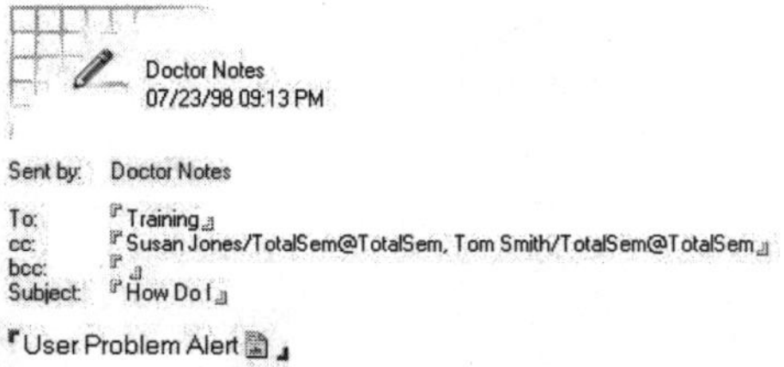

Figure 6.7 Mail message result of @MailSend.

Sample Database

Figure 6.8 is a workflow diagram of the User Question Database. You can add extra steps to the process to handle potential problems. In particular, what can be done to ensure that only one person in the support or training department responds? How can management stay abreast of work being performed? There are many answers to these questions and not all of them involve Lotus Notes!

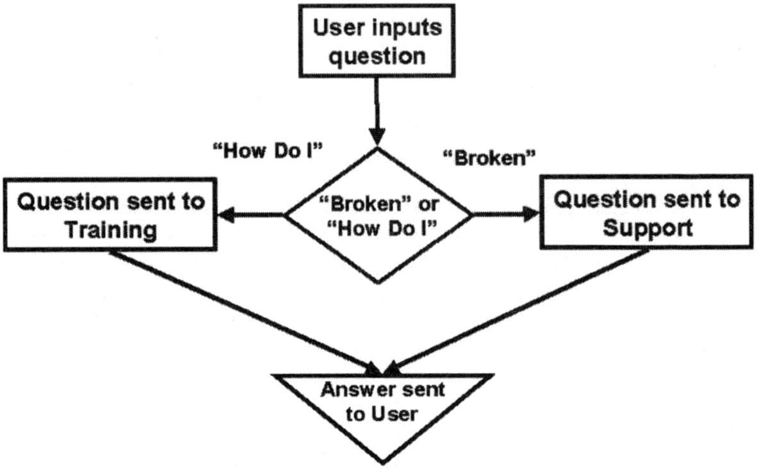

Figure 6.8 Workflow diagram.

To incorporate some of the workflow knowledge learned in this chapter, mail-enable the forms in the User Question database. Add a line to the Close buttons created in the Answers and Questions Forms to automatically send the information along the workflow using the @MailSend function. Send the mail to the Training group you made in the previous chapter for *How Do I* questions and send *Broken* problems to the Support group. CC every email, regardless of category, to the two heads of each group. Figure 6.9 shows what to add to the Close button on the Questions form.

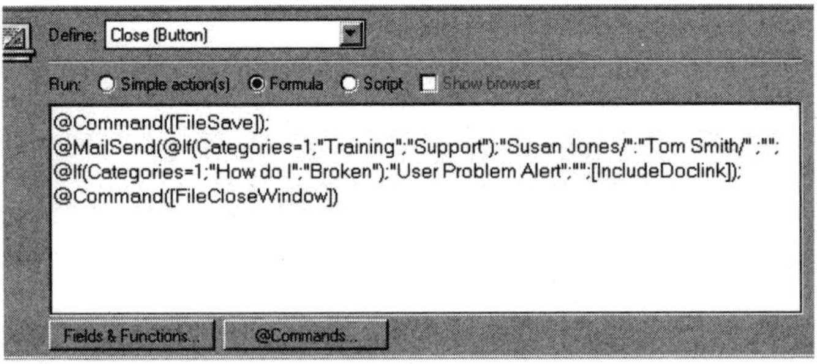

Figure 6.9 Close button formulas.

The @MailSend for the Answer Form is much simpler. We'll leave that for you to figure out. (Hint: Send it to the Uname field!)

Review Questions

1. In creating her first workflow application, Tyene chooses to mail-enable the forms by choosing the option *On Close Present Mail Send Dialog from the Form Properties InfoBox*. Her users complain that they get an error every time they try to save and send the documents. What did Tyene forget to do to the form?

2. Match the term with its definition:

a. Work	1. Activities and Interactions used to reach a business goal
b. Flow	2. Use a process to route forms and data to users
c. Workflow	3. Route links to forms or data to users
d. Business Process	4. Necessary interactions required to reach a business goal
e. Sharing model	5. Activities used to reach a business goal
f. Sending model	6. Rely on users to view forms and data in a database
g. Hybrid model	7. Steps used to reach a business goal

3. After receiving a request for a new database, Leanna begins creating the database immediately on the server. After creating all the forms and views, she will show the database to her users to get their input and to confirm the flow of the business process. This is the best practice process for application development.

   a. True

   b. False

4. What are the three requirements to make a form route correctly? Select three:

   a. Form must be mail-enabled.

   b. Form must have an address on it.

   c. Form must have a route to a mail-enabled database.

   d. Form must exist in a mail-in database.

5. What steps do you have to take to create a mail-in database?

   a. Create a Mail-In database document in the Personal Name and Address Book.

   b. Choose Enable for Mail in the Database Properties InfoBox.

   c. Create a Mail-In database document in the Public Name and Address Book.

   d. Create a Mail-In database document in the database that you want to route mail to.

6. The following formula would do which of these things?

   @MailSend ("Libby Schwarz": "Erick Jones"; "Mike Smith"; ""; "New Restaurant"; "This is the Restaurant Profile for" + RestaurantName; ""; [PriorityHigh]: [IncludeDocLink])

   a. Sends a mail message to Libby, Erick, and Mike using the subject *This is the Restaurant Profile for* that includes the RestaurantName field in the Body of the mail memo. It is sent with High Priority and includes a link to the RestaurantProfile document.

   b. Sends a mail message to Libby and Erick with a copy to Mike. Uses the subject *This is a Restaurant Profile for* and includes the RestaurantName field in the Body of the mail memo. It is sent with High Priority and includes a link to the RestaurantProfile document.

   c. Sends a mail message to Libby and Erick with a copy to Mike. Uses the subject *New Restaurant.* Includes the text *This is the Restaurant Profile* and the RestaurantName as the Remark in the message. Includes a DocLink to the current document when sending and sends the message High Priority.

   d. Sends a mail message to Libby, Erick, and Mike. Uses the subject *New Restaurant.* Includes the text *This is the Restaurant Profile* and the RestaurantName as the Remark in the message. Includes a DocLink to the current document when sending. Sends the message High Priority.

7. Dudley creates a form that he mail-enables by choosing the form options On Close: Present Mail Send Dialog. He also creates a field called MailOptions, sets to the value 1, and an editable field called SendTo. What happens when the document is saved?

   a. The Mail Send dialog is presented. The form is mailed or saved depending on the choices in the Mail Send dialog.

   b. The Mail Send dialog is not presented. The form is mailed.

   c. The Mail Send dialog is not presented. The form is not mailed.

   d. The Mail Send dialog is presented. The form is mailed automatically.

8. Robyn put a MailOptions field on her form and set the value to 1. She hid the field so that users do not see it. How did she hide the field?

9. Besides a Mail-In Database, what other types of databases can receive mail?

   a. Workflow-enabled databases

   b. User's Mail databases

   c. Public Name and Address Book

   d. Personal Name and Address Book

Answers

1. She forgot to create a SendTo field on the document to give the document a destination.
2. a. 5; b. 4 ; c. 1; d. 7; e. 6; f. 2; g. 3.
3. b. False. Before creating a database, she should talk to the users to find out the exact needs and business process. Also, always create the database on your workstation first, not on the server.
4. a, b, and c are correct: The form must be mail-enabled, the form must have an address on it, and the form must have a route to a mail-enabled database.
5. c. Create a Mail-In database document in the Public Name and Address Book.
6. c. Sends a mail message to Libby and Erick with a copy to Mike. Uses the subject *New Restaurant.* Includes the text *This is the Restaurant Profile* and the RestaurantName as the Remark in the message. Includes a DocLink to the current document when sending. Sends the message High Priority.
7. d is correct. The Mail Send dialog is presented. The form is mailed automatically.
8. She enabled the Hide-When options for the field for both reading and editing.
9. b. User's Mail databases.

CHAPTER 7

Advanced Form Elements

The previous chapter described advanced applications, including multi-database applications and workflow-type applications. In addition, the chapter described the advanced application elements necessary for workflow and mailing. Similar to the advanced topics for applications, advanced topics in form creation and form elements are covered on the Application Development II exam. These form elements include subforms, sections, Hide-When options, and layout regions. With the exception of layout regions, each of these topics was discussed for the Application Development I exam. In this chapter, we will briefly review the topics as they were covered for Application Development I and then we will build on that understanding to describe the more advanced details of these elements.

Chapter Objectives

You should be able to answer questions based on the following objectives after reading this chapter:

- Create forms that use subforms.
- Insert subforms using a formula.
- Create forms that use sections to organize information.

- Create controlled access sections.
- Use Hide-When to determine form appearance.
- Use Hide-When for Previewing documents.
- Use Layout regions on forms.
- Compare Layout regions with other form design elements.
- Use other design elements correctly inside Layout regions.

Subforms

As described in the text for the Application Development I exam, *subforms* are database design elements that are created like forms and contain the same elements as forms, such as fields, static text, objects, and sections. Subforms can then be used as a shortcut to create forms. The main purpose of a subform is to allow the designer to reuse design elements in multiple forms, which both makes form building easier and also makes forms more consistent. Also remember that when a subform is edited or changed, all the forms that contain that subform are also changed. This makes editing forms easier for designers. Subforms can be used to create forms in any of the following ways:

- Subforms can be included in the Insert . . . Subform dialog box. This option allows the designer to add the subform to the form they are creating at any point. This option is available on the Subform Properties InfoBox.
- The subform can be included in the New Form . . . dialog box. This option includes the subform in the dialog box that is displayed to designers when they first create a new form. They then have the option to base the new form on the existing subform. The subform appears immediately at create time if this option is selected. This option is available on the Subform Properties InfoBox.
- The subform can be added by a formula. This is called a *Computed Subform.* The designer can add a computed subform by checking the *Insert Subform based on formula* option in the Insert . . . Subform dialog box, shown in Figure 7.1. This option allows the subform to be displayed based on values in the parent form, user roles or names, or other form events.

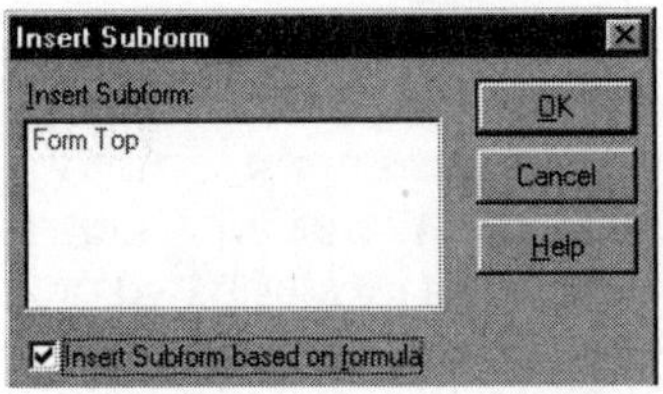

Figure 7.1 Insert Subform dialog box.

EXAM TIP

The process of inserting subforms manually, using the first two options discussed, was described for the Application Development I exam. The main focus of the Application Development II exam, however, is the computed subform.

To use a computed subform, create the subform and choose the option to include the subform in the Insert Subform dialog box. Then, when you create or edit the form on which you want to use the subform, choose Create . . . Insert Subform . . . in the Insert Subform dialog box. Instead of choosing a subform and clicking OK, choose the Insert Subform based on formula option. This displays a computed subform placeholder in the form and allows the designer to create a computed subform formula. The computed subform formula can use field values from the parent document, user roles, or other characteristics that are available at Compose time.

NOTE

The computed subform formula must resolve to the name of a subform contained in the database. If it does not, it generates a run-time error, and no subform is displayed.

One of the most important details about a computed subform formula is that it is only evaluated at Compose or Open time. The computed subform formula is *not* reevaluated if the values of the fields change, for example, or if the user refreshes the fields while in the document. The formula would already have been evaluated, and a subform displayed, at that point.

The computed subform formulas can be created using fields and @functions. This formula will not accept @Commands, simple functions, or LotusScript. An @If function is often used for a computed subform formula, such as:

```
@If (Software = "1"; "Sub1"; "Sub2")
```

This formula says that if the field Software results in the keyword value 1, use the Sub1 subform; otherwise, use the Sub2 subform. If this form is a newly created main document, the subform is inserted based on the default value of the Software field, if there is one. Otherwise, the formula inserts the Sub2 subform automatically, based on a null value in the field at Compose time. An example of when it would be better to use a formula like this would be for a response document that inherited the value of the field Software from its parent document.

Another computed subform formula example would be the following:

```
@If (User = "Libby Schwarz/TotalSem"; "Author"; "Reader")
```

If the field User is computed when composed based on the author of the document, the user Libby Schwarz would be shown the subform Author, and any other user would be shown the subform Reader. Formulas like this one, that refer to User Roles, which will be discussed in detail later, are also useful for computed subform formulas. This type of formula is useful when one subform contains approval or verification information to which only certain users should have access.

The following are the two most important things to remember about computed subforms for the Application Development II exam:

- The computed subform formula must result in a text value that is the name of a subform.
- The computed subform formula is *only* evaluated at compose time or open time and will not be reevaluated based on changes in fields or refreshing the form.

Controlled Access Sections

In the Application Development I exam, you were tested over your ability to create sections within forms to organize data for users. Sections allow the user to collapse or expand areas of the form to hide or display the text, fields, and other objects on that portion of the form. They are collapsed and expanded either manually, by allowing the user to click the twisties displayed next to the section title, or automatically, based on Auto Expand or Collapse options in the Section Properties InfoBox. The main benefit of a standard section is the ability to add design elements and fields to the form without adding length.

There are two main types of sections: standard sections and controlled access sections. The Application Development I exam discusses the process of creating and using standard sections. Application Development II, however, focuses more on creating and using controlled access sections.

Although a controlled access section has most of the same characteristics of a standard section, the controlled access section limits a user's access to edit information. Anyone with access to the database can read the information in the section (unless a Hide-When formula is employed), but only those users defined as editors have the ability to edit the information in the sections.

To create a controlled access section, open the form in design mode and select the text and other design elements that should be in the section. Choose Create . . . Section . . . Controlled Access. Notes displays the Controlled access Section Properties InfoBox. The options on the InfoBox for controlled access sections are slightly different than those for standard sections.

- *Title*. The Title tab allows the designer to give the section a title (see Figure 7.2). Unlike Standard Section titles, Controlled access Section titles cannot be determined by a formula. They must be typed as static text. You can also edit the section field name, the border style, and color on this tab.

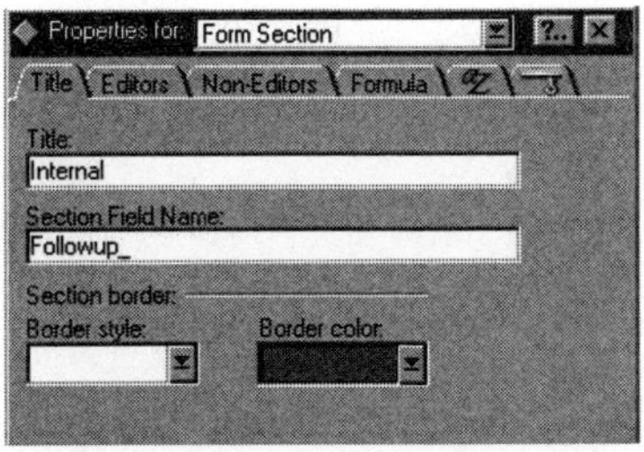

Figure 7.2 Controlled Access Section Properties InfoBox Title tab.

- *Editors*. The Editors tab allows the designer to define the options for expanding and collapsing the section for users that are defined as editors of the section. By default, all the options on this tab are set to Don't auto expand or collapse. This displays the section to the user as it was last viewed. Figure 7.3 shows the *Opened for reading* option set to auto-collapse the section and the *Opened for editing* option set to auto-expand the section. This is for the controlled access section being created for the User Question Database. This is done to force the support personnel to open the form for editing to get all the information about this problem.

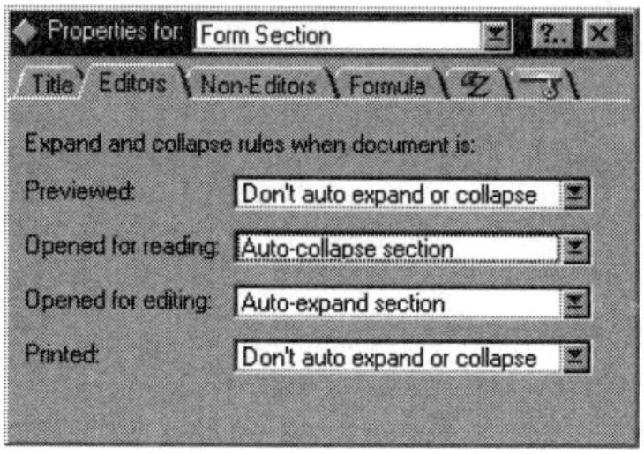

Figure 7.3 Controlled Access Section Properties InfoBox Editors tab.

- *Non-Editors*. The Non-Editors tab allows the designer to define the options for expanding and collapsing the section for users that are not defined as editors of the section (see Figure 7.4). By default, all the options on this tab are set to Don't auto expand or collapse. This displays the section to the user as it was last viewed.

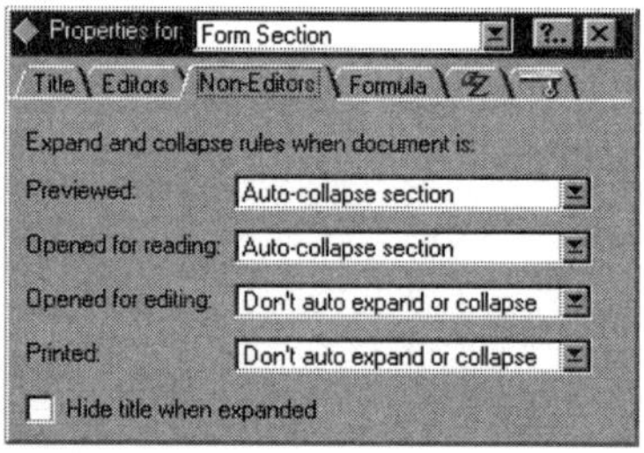

Figure 7.4 Controlled Access Section Properties InfoBox Non-Editors tab.

- *Formula*. The Formula tab, shown in Figure 7.5, allows the designer to define the editors for the section. The section editors formula can be editable, computed, computed for display, or computed when composed. If the formula is editable, access to the section can be changed at a later time, whereas if it is computed, it cannot be changed outside of the design phase. A computed when composed formula creates a permanent editors list at compose time.

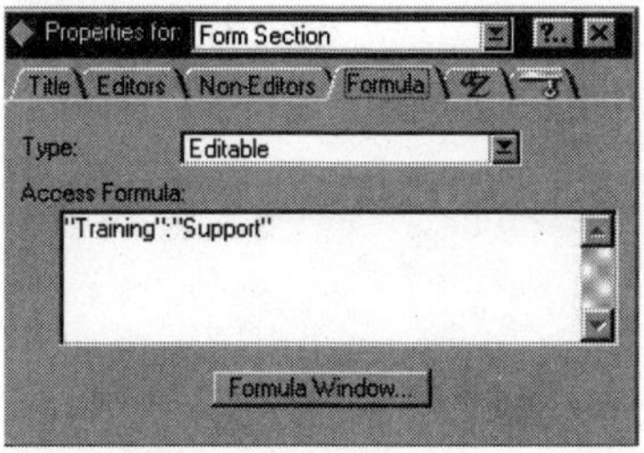

Figure 7.5 Controlled Access Section Properties InfoBox Formula tab.

Formulas to define the section editors often make use of user names or Access Control List roles. These formulas must resolve to text lists containing one or more names. Roles are discussed in Chapter 11. The following is an example of a formula that might be used to control editor access:

"Libby Schwarz/TotalSem": "Marketing Group"

This formula allows Libby and the group in the NAB called the Marketing Group to edit the section.

You can also use a formula that resolves to @UserNames. If you choose to use this as the formula, anyone with Edit access to the document is given the ability to add editors to the section by selecting Section . . . Define Editors. . . . This menu option is only available when the document is in edit mode and the section in the document is collapsed. This menu option displays the Edit Section dialog box, shown in Figure 7.6. The author would then be able to assign the ability to edit the section to other users.

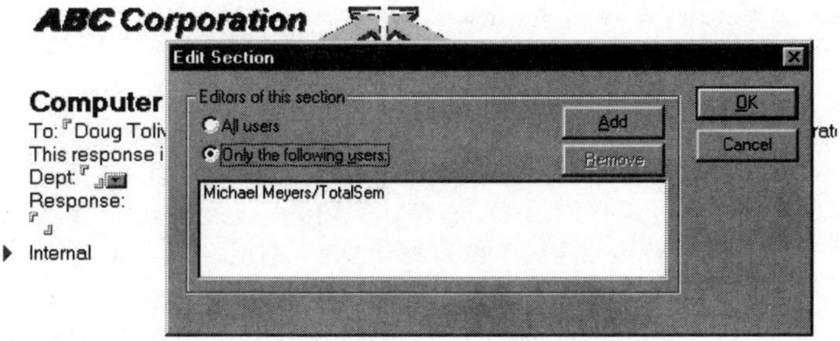

Figure 7.6 Edit Section dialog box.

NOTE
This formula cannot change the Access Control List for the database. If the users listed in this formula as editors do not already have editor access to the database, they will not be given the ability to edit the document through this formula.

- *Text.* The Text tab, like the Text tab for the Standard Section, allows the designer to define the appearance of the section title.
- *Hide-When.* The Hide-When tab is the same as the Hide-When tab for text, objects, and standard sections. The designer can use this tab to hide the section based on a form event or on a formula. The Hide-When options will be discussed later in the chapter.

The following are the main focus points about controlled access sections:

- The designer can limit the ability to edit the section based on a formula.
- The section can be expanded or collapsed based on the defined editors.
- The controlled access section formula does not and cannot override the Access Control List.
- Users with reader access to the database will be able to see the data in a controlled access section, even though they cannot edit it.

NOTE
The security aspects of Controlled Access sections, like most security functions, do not work on a local (workstation) database. To test the functionality, either place the database on a server or open the Access control dialog box to the Advanced panel and enable the Enforce Consistent ACL option.

Using Hide-When Options

Hide-When options allow designers to hide design elements based on a document event or a formula. Hide-When was described and tested in detail in Application Development I. Hide-When options can be enabled for most design elements including buttons, hotspots, graphics, text, sections, and other objects. To enable Hide-When options, open the Properties InfoBox for the selected object. Click the Hide-When tab (looks like a window shade). On the Hide-When tab, you can choose to hide elements based on predefined document events including Previewed for reading, Previewed for editing, Opened for reading, Opened for Editing, Printed, and Copied to the Clipboard. For example, to hide the directions for completing a form from anyone only previewing the document, select the paragraph and choose Previewed for reading and Previewed for Editing from the Hide-When options. In this example, you would also most likely choose to hide the paragraph when Opened for reading.

NOTE
In Release 4.6, designers are also given the option to hide a design element from Web browsers and from Notes 4.6 or later. These options will not be tested on the exam, but may help in understanding the options, especially if you are learning by using Notes 4.6.

The Hide-When tab also gives the designer the ability to hide elements based on a formula. This formula can be based on the value of a field, a user name or role, or other function. The following is an example of a Hide-When formula that would hide the section based on a field value:

```
Categories != Software
```

In this example, the designer could hide a section of the form based on the value of a keyword field called Categories. If the value of the field Categories is not equal to Software, the section covered by the Hide-When formula would be hidden. If the value of the Categories field is Software, the paragraph is displayed to the user.

NOTE
If the value that the Hide-When formula is based on is a keyword field selection and can be changed by the user or recalculated during compose time, the field on which the formula is based must have the option "Refresh fields on keyword change" enabled.

Layout Regions

Layout regions are new design elements for Notes Release 4. For this reason, they are tested heavily on the Application Development II exam. The main purpose of a layout region is to give the designer more control over the placement of text and other design elements. In a layout region, the designer has the ability to place items precisely in the form. In a form, the designer normally relies on alignment and spacing to place design elements in a particular location. In a layout region, however, the designer can drag elements around a grid to place them at a precise location in the region.

One specific benefit of a layout region is the ability to layer design elements on top of one another, allowing more visually interesting elements as well as allowing the designer to control field tab order. To layer a button or text on top of a graphic, for example, place the elements in the layout region and then shuffle them using the Bring to front and Send to back options. Layering is especially useful because of the way layout regions treat field order. In a form, fields are accessed left to right, top to bottom, when a user tries to tab through them. In a layout region, by comparison, the fields are accessed in the order in which they were created. To avoid this, use layering to set the tab order by sending fields to back or bringing fields to front. First, choose the field that should have default focus and select bring to front. Find each field in order and click bring to front.

Layout regions do not support all the design elements that forms and subforms support. Table 7.1 compares the supported and unsupported elements in a layout region. Layout region supported and unsupported elements are tested heavily on the Application Development II exam.

Table 7.1 Element Support in Layout Regions

Unsupported Elements	Supported Elements
Sections	Static text
DocLinks, Database Links, View Links	Graphics
Rich text fields	Graphic buttons
Dialog list type keyword fields	Buttons
Hotspots, including pop-ups, except for buttons and graphic buttons	Fields (except Rich Text fields and Dialog list keyword fields)
Attachments	Layering
Tables	Listbox and Combobox type keyword fields

NOTE
Layering and Listbox and Combobox type keyword fields are not supported outside of layout regions.

To create a layout region on a form or subform, place the cursor on the form and choose Create . . . Layout Region . . . New Layout Region from the menus. This inserts a layout region box that can then be sized. To change the size or other characteristics of the layout region, right-click and select Layout properties to display the Layout Properties InfoBox, shown in Figure 7.7.

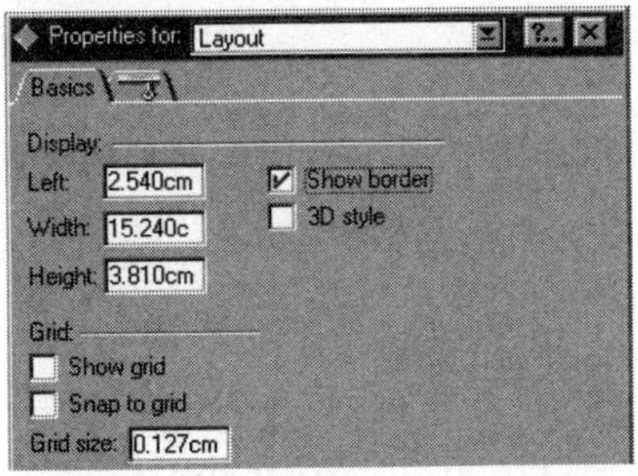

Figure 7.7 Layout Region Properties InfoBox.

The Basics tab on the InfoBox allows the designer to determine the size and position of the layout region using the Left, Width, and Height fields. In addition, the layout region can have a border or a three-dimensional style gray background using the Show border option and the 3D-style option.

The other options on the Basics tab allow the designer to display a grid. The grid allows the designer to line up elements in the layout region with each other more easily. There are also grid size options and snap to grid options. The other tab on the Layout Region Properties InfoBox is a Hide-When tab. Layout regions can be hidden using the same document events and formulas that are used to hide other design elements.

After creating a Layout Region, the designer can place other design elements in it, based on the supported elements, as described. To place static text in a layout region, either click the Create Textbox SmartIcon or choose Create . . . Layout Region . . . Text. Notes displays an untitled text box in the layout region. Double-click the text box to display the *Control* InfoBox. In layout regions, instead of text and other elements having properties, they have controls. These controls are defined by the designer for each element using the Control InfoBox in a similar manner to a Properties InfoBox. A static text Control InfoBox is shown in Figure 7.8.

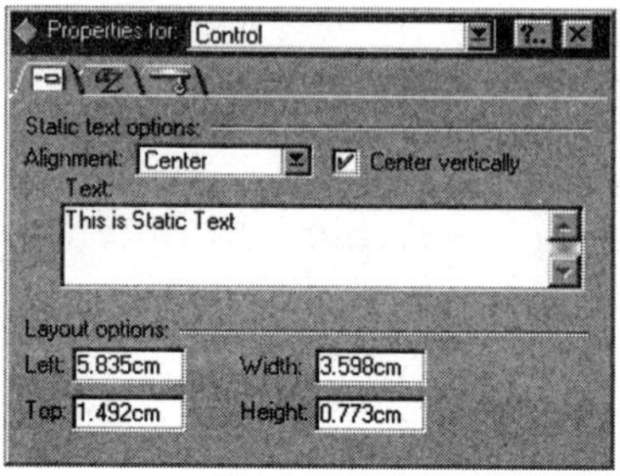

Figure 7.8 Static Text Control InfoBox.

The first tab allows the designer to set the alignment of the text (left, right, or center) as well as whether or not the text should be vertically aligned. The text itself is defined in the Text: field on this tab. Type in any text that you want to appear in this text box in the layout region. Finally, you can use the Left, Top, Width, and Height options to set the size and position of the text box. The size and position can also be set by clicking and dragging on the control boxes at the edges of the text box itself.

The Text tab, next in the Control InfoBox, allows the designer to set the text properties for the text in the text box, including font, size, color, and style. Finally, the last tab in the Control InfoBox is a Hide-When tab. Hide-When in layout regions is different than in a form or subform. One of the defining characteristics of Hide-When is that it is paragraph-level formatting, meaning that if you enable a Hide-When formula or option for any part of a paragraph (up until you press Enter), those same Hide-When options are enforced for everything in that paragraph or on that line. In layout regions, however, each element can be hidden independently. You can place a graphic, a button, and some text on a single line, for example, and have three different Hide-When formulas for the three elements.

NOTE

You cannot type text in a layout region the way you would when creating a form or subform, nor can you copy and paste text directly into a layout region. You must use text boxes instead, as described above.

Another element that you may want to add to a layout region is a field. Fields are created in layout regions in the same way they are created on forms or subforms. In layout regions, however, fields have some additional characteristics defined by their controls.

When you add a field to a layout region, it is displayed as an Untitled Edit Control.

All field types except rich text fields and dialog list keyword fields can be used in layout regions. For all field types except for keyword lists, the following additional options are available, accessed on the Control tab of the Field Properties InfoBox:

- Edit Control Options, including multiline, scrollbar, and border.
- Layout options, including left, top, and width.

These options allow the designer to determine the size, appearance, and position of the text, time, number, authors, readers, and names types of fields when these are placed within a layout region. These controls allow a designer, for example, to require a text field to have multiple lines. It is best to use the multiline and scrollbar options together to give the cleanest look to the user. The text field options that are available in a layout region are shown in Figure 7.9.

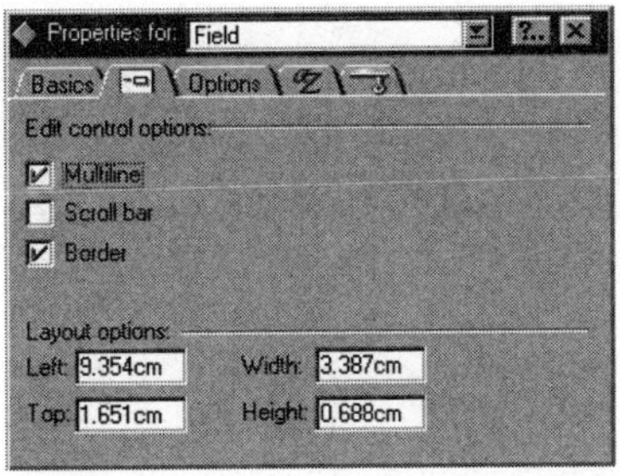

Figure 7.9 Text field properties in a layout region.

NOTE

One special option that is not tested but that you may find useful is the ability to display calendar and time pop-up choices when using editable time fields in layout regions. These options are similar to the ones in the Notes Mail database. To see this display, create a field in the layout region and make it an editable time type of field. To display the calendar pop-up, choose to display date only. To display the time pop-up, choose to display time only. The result of the calendar pop-up is displayed in Figure 7.10.

The interfaces available for keyword fields in a layout region differ from those in other regions. First, a dialog list type of keyword field is *not* available in a layout region. Instead, layout regions use two additional keyword field types, list boxes and combo boxes.

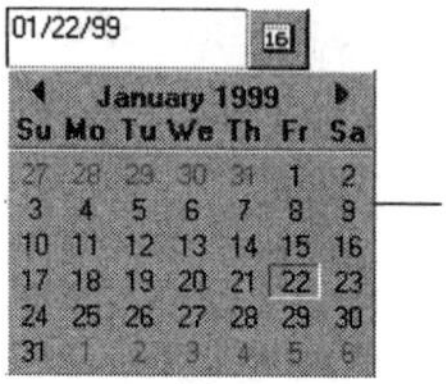

Figure 7.10 Calendar pop-up using Date fields in a Layout region.

A *list box,* as shown in Figure 7.11, allows the keyword options to be displayed in a list. The number of options displayed at a time depends on the size of the field in the layout region. If there are more options than can be displayed in the field, the field displays a scrollbar. List boxes allow users to select multiple values by placing a checkmark in the list box if the Allow multi-values option is enabled.

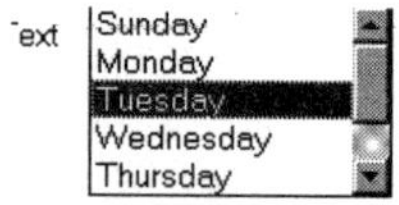

Figure 7.11 List box.

A *combo box,* as shown in Figure 7.12, allows the keyword options to be displayed using a drop-down list. The first keyword option is displayed by default, unless a default value formula is created, and can be changed by clicking the drop-down arrow at the right side of the field. Note that the Allow multi-values option is not available for the combo box interface.

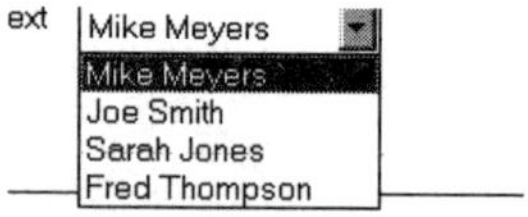

Figure 7.12 Combo box.

Keyword fields in layout regions also have no option to *Allow values not in list.* Note as well that only the values provided by the designer can be used in keyword fields in a layout region.

NOTE

Be careful when selecting fields from a layout region for placement on other forms or subforms. When you copy a field from a layout region, you copy not only the field but also the entire lay-

out region. You can copy and paste fields into layout regions, but you may not want to copy fields from layout regions and paste them directly into forms.

One important use of layout regions is creating dialog boxes. You can combine the @Dialog box function (discussed in Chapter 8) with a form created with a layout region. This creates an attractive, user-friendly dialog box.

One of the tricks on the exam is to recognize that each of the form elements described in this chapter has some similar functionality and some very different functionality. The elements can sometimes be combined to make developing an application easier. To reuse controlled access section elements, for example, first create them on a subform. Then insert the subform anywhere this section is needed. Table 7.2 compares a few salient points about subforms, sections, and layout regions.

Table 7.2 Comparison of Subforms, Sections, and Layout Regions

Subforms	Sections	Layout Regions
Can have several on a form	Can have several on a form	Can have several on a form and can take up the entire form
Can use formulas to determine when it is displayed, using a computed subform	Can be hidden based on Hide-When options or formulas and editing can be controlled by a formula	Can be hidden based on Hide-When options or formulas
Can place sections or layout regions on subforms, to reuse them	Cannot use a section in a layout region	Layout regions cannot display sections or subforms
Used to create customized forms	Used on forms to organize information	Used on forms or subforms to create visually interesting regions

Sample Database

Both the Training and the Support departments respond to questions, but they want to do a little internal advertising to separate themselves from each other. To do this, they have each made special company logos reflecting their departments. They want users who are reading their Answer forms to see their new logos at the top of the form instead of the standard company logo. Plus, they want the Training logo to appear if the answer comes from training and the Support logo to appear when the answer comes from support. We will use the Formula-based subform feature to perform this, so follow these steps:

1. Create two new subforms called Ftraining and Fsupport. You can copy the Form Top subform using standard copy and paste, and then rename the subform. Create two new logos for each department. Replace the company logo with the training logo in the Ftraining subform and the support logo in the Ftraining subform. Figure 7.13 shows an example of the Fsupport subform. Don't delete Form Top! It is still needed for the Question form.

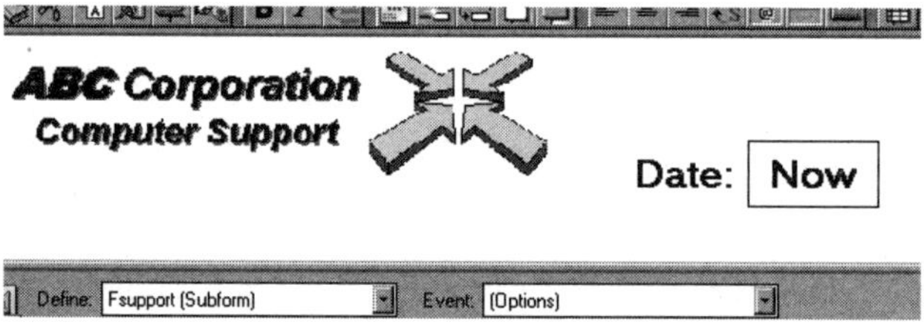

Figure 7.13 Sample subform.

2. Open the Answers form in design mode and delete Form Top. Go to Create/Insert Subform . . . and click the *Insert Subform based on formula* checkbox. It should look like Figure 7.14.

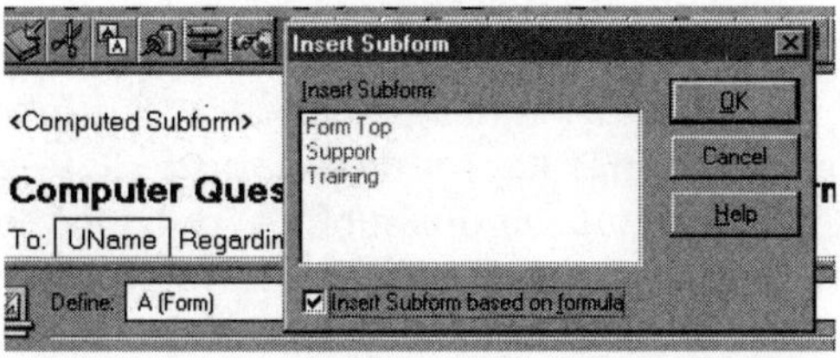

Figure 7.14 Inserting a formula-based subform.

3. Add the formula as shown in Figure 7.15 to finish the job.

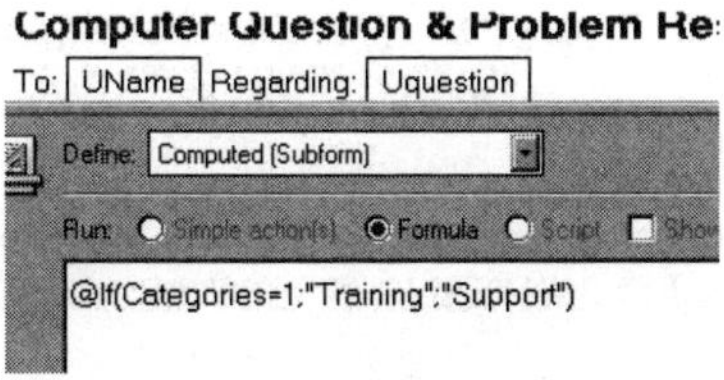

Figure 7.15 Finished formula-based subform.

Layout regions can sure make the subforms look better. Use layout regions to modify the logo and the Date field so the Date field is vertically centered with the logo.

Another problem with the Answer form is the use of Hide-When for the fields used by support personnel. In the past, certain management users had edit rights that allowed them to see the sometimes negative comments the support people wrote about them. To avoid this in the future, we will change those Hide-When fields into a controlled access section. Simply select all the static text and fields that will be included in the Controlled Access Section and select the Design . . . Section . . . Controlled Access Section. Then add the Training and Support groups the Editable section in the Properties . . . Formula tab. If you will refer all the way back to Figure 7.5, this shows how it should look.

Review Questions

1. Jeanine creates a set of subforms that she chooses to display based on the value of the Ratings keyword field. If the Ratings field is set equal to 1, the Congratulations subform is displayed. If the Ratings keyword field is set equal to 2, the Improvements subform is displayed. Which subform is displayed if the default value of the Ratings field is 1, but in the course of editing the document you change the value to 2?

   a. Congratulations

   b. Improvements

2. Teresa creates two subforms that she uses when creating response documents in her Term Paper Ideas database. She bases the subform formula on the value of a field in the main document. Is this a good use of a subform and will the correct subform be displayed to the users if the formula is : @If (Status = "Approved"; "GreatIdea"; "TryThisIdea")?

3. A controlled access section will prevent users not in the Editors list from seeing the data in the section.

   a. True

   b. False

4. The title of a controlled access section can be determined by a formula, just like the title of a standard section.

   a. True

   b. False

5. Maggie created a form with a controlled access section that uses a computed formula to define the editors. Can the author of a document created with this form add an editor to the section?

   a. Yes

   b. No

6 . When Carolyn's users open the Business Card Order form, some of them see the Approval section expanded and some of them see the Approval section collapsed. How did Carolyn accomplish this?

a. Editors of a section automatically see the section expanded.

b. Non-Editors of a section automatically see the section expanded.

c. Carolyn chose for editors to see the section expanded and non-editors to see the section collapsed.

d. You can't control whether a section is expanded or collapsed. It automatically reverts to the way it was last displayed.

7. Henry is defined as a reader of the Purchase Order database in the Access Control List. The designer put Henry in the Editor formula of a controlled access section. Can Henry edit the fields in the controlled access section?

a. Yes

b. No

8. Pamela's users often use the Preview function of Notes to scan the documents in the Company News database. When they are scanning the documents, they want to be able to see the main body of the document without having to scroll. The top half of the document usually has some header information such as date, author, and topic. Which of the following options should Pamela use to ensure the users get what they want?

a. Put all the header information in a standard section. Use an Auto Collapse when Previewed option to limit the entire section to one line when the document is previewed.

b. Select all the header information and open the Text Properties InfoBox. Use the Hide-When tab to hide the text when the document is previewed for reading or editing.

c. Put all the header information in a standard section. Use the Hide-When options for sections to hide the text when the document is previewed for reading or editing.

d. Use the Form properties to enable the On Preview Show Body Only option.

9. Write a formula that would hide text, based on the value of the Status field. If the Status is not Approved (keyword synonym of 1), the text should be hidden.

10. Mike doesn't like the dialog list format of a keyword type field. He prefers the idea of a combo-box. How can he create a combo-box for keywords instead of a dialog list?

    a. Choose the Combo-Box Interface instead of the Dialog list interface.

    b. There is no Combo-Box Interface.

    c. The Combo-Box Interface is available if the keyword field is in a layout region.

    d. The Combo-Box Interface is the default.

11. Choose the elements that can be used in a layout region:

    a. Sections

    b. Graphics

    c. Rich Text fields

    d. Radio button interfaces for keyword fields

    e. Tables

    f. Dialog list interfaces for keyword fields

    g. Attachments

12. Which of the following is a benefit of using a layout region?

    a. You can place text more precisely.

    b. You can allow users to enter values not in list in the keyword fields.

    c. Users can tab through fields left to right and top to bottom.

    d. You can make a layout region into a dialog box.

13. You can have several layout regions on a form.

    a. True

    b. False

14. You can have several sections on a form.

    a. True

    b. False

15. You can add a layout region to a subform or a subform to a layout region.

    a. True

    b. False

16. You cut and paste a field that was in a layout region onto a subform. What happens?

17. Which of the following keyword options are not available when the keyword field is used in a layout region:

    a. Dialog List

    b. Radio Button

    c. Allow Multi-Values

    d. Allow Values Not In List

Answers

1. a. Congratulations will always be displayed in this case because subform formulas are only evaluated at compose time.

2. Yes; the formula will show either the GreatIdea subform or the TryThisIdea subform when the response form is created, based on the value of the Status field on the main document.

3. b. False. Users can still see the data if they have reader access to the database. The Controlled access section prevents them from editing the data in that section.

4. b. False. Controlled access section titles must be static text.

5. b. No. The editors can only be changed by changing the formula in design mode.

6. c. Carolyn chose for editors to see the section expanded and non-editors to see the section collapsed.

7. b. No. The editor formula of a controlled access section does not override the ACL.

8. Options a, b, and c would all accomplish the desired look when the users are previewing the document.

9. Place the formula Status != 1 in the Hide-When formula.

10. c. The Combo-Box Interface is available if the keyword field is in a layout region.

11. b and d are supported in Layout Regions; a, c, e, f, and g are not supported in layout regions

12. a. You can place text more precisely.

13. a. True

14. a. True

15. b. False. You can place a layout region on a subform, but you cannot place a subform on a layout region.

16. The entire layout region gets pasted with the field.

17. a and d. You cannot use the dialog list interface and you cannot choose to allow values not in list when the keyword field is in a layout region.

CHAPTER 8

Advanced Fields, Functions, Formulas, Actions, and Agents

In the previous section, we reviewed some basic form concepts such as subforms, sections, and Hide-when options, and then built on those by creating computed subforms and controlled access sections—as well as the new design element of layout regions. As we make the databases and applications more advanced, you will want to use more advanced fields, functions, formulas, actions, and agents. These enable you to add more advanced functions to your applications.

For example, in Chapter 6, *Advanced Applications and Workflow Database Design*, we discussed the advanced possibilities with mail functions, such as the MailOptions reserved field and the @MailSend mailing function. This chapter will build on similar concepts by discussing field types, including other reserved fields and internal fields, and advanced functions. In addition, we will discuss the advanced functions possible with multi-line formulas and advanced actions and agents.

Chapter Objectives

After reading this chapter, you should be able to answer questions based on the following objectives:

- Use functions that interact with the NOTES.INI file to store and retrieve data.
- Use reserved fields for mailing, signing, and saving documents.
- Use Notes internal fields to find information about or troubleshoot a document.
- Use functions to interact with other databases to obtain information.
- Present the user with dialog boxes to obtain information.
- Use formulas to complete the choices available in keyword fields.
- Use multi-line formulas.
- Use functions to stop other functions from continuing to evaluate.
- Display forms in dialog boxes.
- Use functions to manipulate lists of data.
- Use functions to manipulate user and server names.
- Use scheduled/automatic agents.
- Test and troubleshooting agents.
- Create actions that run agents.

Field Types

For the Notes Application Development I exam, we discussed the basic field types available in Notes: Editable and Computed. The editable fields have events for which formulas could be written: default value, input translation, and input validation. The computed fields can be broken down into three types: computed, computed for display, and computed when composed.

Editable fields are fields in which the user can enter or change data. The default value event enables the designer to specify an initial value for the field, to be displayed for the user at compose time. The input translation event contains formulas that edit the values entered by the user. For example, if a designer wants to ensure that the names and addresses the user enters are in proper case and do not contain extra spaces, the designer will use an @ProperCase and an @Trim function in the input translation event. The input validation event enables the designer to require completion of certain fields. If the designer wants to ensure that the name and address fields are completed before the document is saved, the designer will use an @If function with @Success and @Failure in the input validation event.

Computed fields are used to calculate values for the user or internally in the application. The basic computed field must have a formula to provide the value, as users cannot edit the data in a computed field. A *computed for display* field is calculated every time the user views, refreshes, saves, or loads the field, but it is never saved with the document. A computed for display field should be used when the designer wants to display information to the user while the user views the document—but when the user does not need that information for other calculations or views. A *computed when composed* field is calculated only once, at compose time. This type of computed field is useful for a time/date stamp or a field that captures the author's name, for example.

One of the concepts tested on the Application Development I exam that you will need to review for Application Development II is recognizing when the formulas placed in these fields are evaluated. Table 8.1 reviews this concept.

Table 8.1 When fields are evaluated

Field Type	When it is Evaluated
Editable, Default Value	Compose Only
Editable, Input Translation	Refresh, Save
Editable, Input Validation	Refresh, Save
Computed	Compose, Refresh, Save
Computed for Display	Compose, Refresh, Save, Load
Computed When Composed	Compose Only

NOTE

Remember that *computed for display* fields are never stored in the database and can never be used in views.

Designers of the Application Development II exam also assume you will have some knowledge of additional field types. We will discuss internal and reserved field types first.

INTERNAL FIELDS

All internal fields in Notes track information that Notes generates automatically. The names of these fields generally begin with a dollar sign ($). To view these fields, open the Document Properties InfoBox, shown in Figure 8.1, and click the Fields tab.

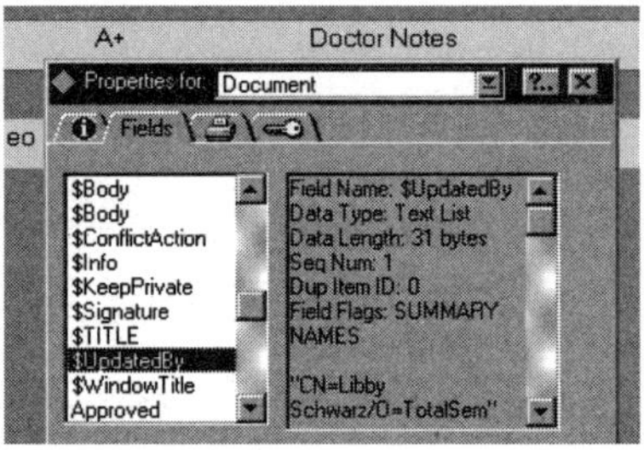

Figure 8.1 Document Properties InfoBox, Fields tab.

Internal fields can be used in Notes formulas but should not be changed, because changing them can corrupt the document. Table 8.2 describes some, but not all, of the important internal fields.

Table 8.2 Notes internal fields

Field	Value and Information
$Title	This field is used when a form is stored in the document. If a form is stored in the document, this field contains the name of the form.
$WindowTitle	This field is used when a form is stored in the document. This function describes the result of the formula used to define the window title.
$Body	This field is used when a form is stored in the document and displays the static text in the form.
$Ref	This field is used with response-type forms. The field stores the parent document's *Unique Notes Identification Number* (UNID) to help Notes maintain the relationship between parent and child documents.

Field	Value and Information
$Revisions	This field tracks the time and date when the document is modified. This field is useful when tracking edits to the document.
$UpdatedBy	This field tracks usernames of people who have edited the document.
$File	This field displays an entry describing each attachment to the document.
$Links	This field displays an entry describing each document, view, or database link in the document.
Form	This field, though named without a dollar sign, is an internal field that displays the name of the form used to create the document—or, if different, the name of the form that was most recently used to save the document.

Each of these internal fields can be used when troubleshooting or gathering information about a document. They also can be used in formulas (in views) or in other forms to gather information.

The *form field* is added to each document that is created, to define the form that was used to create the document. This field is used to define the display form when the document is displayed. As mentioned before, this field can be changed if a user with editor rights to the document changes the form.

Notes can use the form field in two other ways:

- By using the Form field to select only documents created with a particular form to be displayed in a view.
- By using the Form field to define how to display data.

As discussed for the Application Development I exam, sometimes —but not always—the form used to create the document is also used to display the document. If you choose to enable the Store Form in Document option for the form, the form used to create the document will always be used to display the document. The form and the data will merge, and you cannot choose to display the document using another form. If, however, you do not choose to store the form with the document, you can display the data using different forms. This function may be useful in cases where the data necessary for input is not necessary for reading the document—or when you need to change the data format.

As discussed earlier, Notes chooses which form to use in displaying data, based on the following criteria (in order):

1. Notes will first display the data using the form stored in the document, if there is one.
2. If no form is stored with the document, Notes displays the data using the form specified in the form formula. (See Chapter 9, *Advanced Views*, for more information on form formulas).
3. If there is not a form formula specifying the form to use, the data is displayed using the form with which the document was created, if it is available
4. If the form with which the document was created is not available, the default form for the database is used to display the data.

If a user wishes to view the data with any other form while the document is open, select View . . . Switch Form. This option displays the Switch form dialog box, shown in Figure 8.2, and enables the user to choose another form from the database with which to display the data.

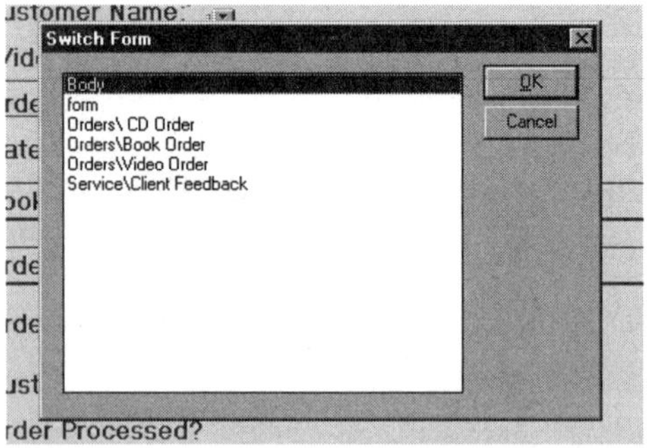

Figure 8.2 Switch form dialog box.

NOTE

Remember that while another form may be available in this list, it may or may not have appropriate fields necessary to display the data. If you want this option to be useful to your users, create similar or descriptive names for forms that use the same fields.

TROUBLESHOOTING TIP

Here is one troubleshooting tip for a problem that can be especially frustrating for users and developers. Sometimes users will complain that the document they worked on is no longer available in the view, but they are sure that they did not delete it. A possible explanation is that the user changed forms while editing the document, and the document no longer meets the selection criteria for the view. The Main view in a database may, for example, use the following selection formula: `SELECT Form = "Book Order"`**. Then, when the user edited a document created with the Book Order form, the user changed the form to the Video Order form because the customer changed the order. Unfortunately, the user did not realize that the document was no longer eligible to be displayed in the Main view.**

Reserved Fields

Reserved fields are predefined fields that can add automatic functions to your forms and documents. Reserved fields provide functions to your forms that you would otherwise have to program yourself, using @Functions, @Commands, or LotusScript. Notes uses many of the reserved fields for mail functions, which were described in Chapter 6 (*Advanced Applications and Workflow Database Design*). To review these reserved fields, see Table 8.3.

Table 8.3 Notes reserved fields for mailing

Field	Value and Information
MailOptions	This field enables the designer to force the document to be mailed at save time. If the MailOptions field value is set to one, the document will be sent automatically. If the MailOptions field value is set to zero, the document will not be mailed. This field takes precedence over the values selected by the user in the Mail send dialog box.

continues

Table 8.3 Continued

Field	Value and Information
Sign	This field enables the designer to force the document to be signed when it is mailed. If the Sign field value is set to one, the document will be signed automatically when it is sent. If the Sign field value is set to zero, the document will not be signed. This field takes precedence over the values selected by the user in the Mail send dialog box. Signing will be defined and described further in Chapter 11, *Security in Applications.*
Encrypt	This field enables the designer to force the document to be encrypted when it is mailed. If the Encrypt field value is set to one, the document will be encrypted automatically when it is sent. If the Encrypt field value is set to zero, the document will not be encrypted. This field takes precedence over the values selected by the user in the Mail send dialog box. Encrypting will be defined and described further in Chapter 11, *Security in Applications.*
SendTo	This field enables the designer to define which users will be sent the document. If the form is mail-enabled, this field is required.
CopyTo	This field enables the designer to define which users will receive a copy of the document.
BlindCopyTo	This field enables the designer to define which users will be sent a blind copy of the document.
Delivery Priority	This field enables the designer to specify the document's delivery priority. It can be set to high, normal, or low, and defines how quickly the message will be passed to other mail servers.
ReturnReceipt	This field enables the designer to specify whether the sender is notified when the recipient opens the document.

Field	Value and Information
DeliveryPriority	This field enables the designer to specify when a delivery report is returned to the sender. The default only sends the delivery report on failure of the delivery. The other choices include Confirm Delivery, Trace Entire Path, and None.
SaveOptions	This field enables the designer to force the document to be saved at the time when it is mailed. If the SaveOptions field value is set to one, the document will be saved automatically when it is sent. If the SaveOptions field value is set to zero, the document will not be saved. This field takes precedence over the values selected by the user in the Mail send dialog box.

Reserved fields also can be used on documents that are not being mailed and are described in Table 8.4.

Table 8.4 Additional Notes reserved fields

Field	Value and Information
Categories	This reserved field name, discussed for the Application Development I exam, enables documents to be categorized in Notes views.
$VersionOpt	This field enables designers to automate version control for documents without using the form properties. This field can be more flexible than Form Properties, as it can be used on a document-by-document basis rather than being applied to all documents created with the form. The field can contain the following values, either as a computed field or as an editable keyword field (the values are shown as they would be used in a keyword field with synonyms):

continues

Table 8.4 Continued

Field	Value and Information
	■ Don't track versions | 0 ■ Create response if File - Save As New Version is used | 1 ■ Create response automatically | 2 ■ Promote to main document if File - Save As New Version is used | 3 ■ Promote to main document automatically | 4 ■ Create additional main document if File - Save As New Version is used | 5 ■ Create additional main document automatically | 6
FolderOptions	This field enables the designer to automate moving documents to folders. The menu option for this is Actions . . . Move to Folder. If the FolderOptions field value is set to one, Notes prompts the user with the Add to Folder dialog box, shown in Figure 8.3, when the document is saved. If the FolderOptions field value is set to two, the document will be placed in the current folder automatically.
SecretEncryptionKeys	To encrypt a mailed document, you use the Encrypt reserved field. To encrypt a document that is not mailed, use the SecretEncryptionKeys field. This field encrypts the document using secret keys rather than the public keys stored in the Public Name and Address Book. Encryption will be described in more detail in Chapter 11, *Security in Applications.*

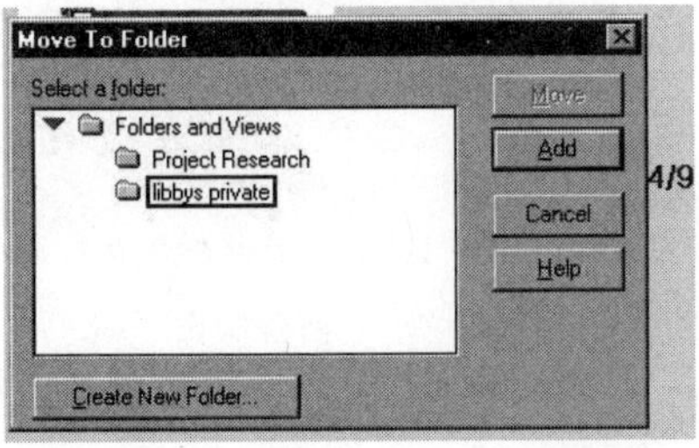

Figure 8.3 Add to Folder dialog box.

Fields in Notes can be of multiple types, including editable and computed. In addition, there are internal and reserved field types, predefined by Notes either to display internal data or to perform a predetermined function. For the exam, it is important to know how the reserved and internal fields can be used and how Notes will evaluate them.

Name Fields and Functions

Another field type that can be used and defined internally by Notes is a Names data type field. This option includes the Names, Authors, and Readers field types. The Authors and Readers field types will be described in more detail in Chapter 11, *Security in Applications*. The purpose of a Names type field is to reference the fully distinguished hierarchical names of users. A *hierarchical name* uses a series of components to identify the user, including the *Common Name* (CN), *Organizational Unit* (OU), *Organization* (O), and *Country* (C). Internally, Notes stores and uses these hierarchical names in the full canonical format, such as the following example:

```
CN=Libby Schwarz/OU=Authors/O=TotalSem
```

Sometimes, however, this format is too cumbersome to display for users. Names fields, therefore, display names in the abbreviated format:

```
Libby Schwarz/Authors/TotalSem
```

The Names field is used when you need to display the user or server name without using the name to assign rights to documents.

The Names field type automatically converts the hierarchical names into abbreviated format for display and to canonical format for internal use. If user or server names are used in other field types and in views, however, the following functions can assist with both returning the name and manipulating the display and storage format of the names. The @Functions listed in Table 8.5 work specifically with user and server hierarchical names:

Table 8.5 Name @Functions

@Function and Syntax	Notes
@Name ([action]; name)	This function enables the designer to manipulate the hierarchical name of a user or server. This function is especially useful for formatting names to be displayed in views. The action parameter enables the designer to expand, abbreviate, or separate the name listed in the function. The actions allowed on a name include the following: ■ [Abbreviate]—Removes the component labels from a hierarchical name. ■ [CN], [G], [I], [S], [Q], [OUn], [O], [C] —Displays the specified component of a hierarchical name. ■ [Canonicalize]—Adds the component labels to a hierarchical name, including adding any missing components. ■ [ToKeyword]—Puts the hierarchical name in a format that will allow it to be categorized in a view.
@UserName	This function displays the current user or server name in canonical format, if the name is hierarchical.

Advanced @Functions

For the Application Development I exam, you studied and mastered the use and syntax of @Functions in Notes. @Functions are the predefined actions that enable designers to create formulas in Notes without having to know an entire programming language, such as C++ or Visual Basic—although that type of function is provided through LotusScript for more advanced developers. For the Application Development II exam, it is still important to understand correct syntax, placement, and evaluation of Notes @Functions. Application Development II also expects people taking the exam to know some of the more advanced @Functions, including functions that use Document IDs, manipulate lists, manipulate the Notes environment, obtain information from other databases or views, and functions that prompt the user for information. These and other advanced @Functions are described in the following sections.

NOTE

Not all @Functions tested on the Application Development II exam are listed and described. A more complete list of Level II @Functions can be found in Appendix B, and the full list of possible @Functions is available in the Notes Help database or in the Lotus Notes Official documentation *Programmer's Guide, Part 2.*

Document IDs

Internal and reserved fields, discussed in the previous section, are created or predetermined automatically by Notes. Another item created automatically by Notes, similar to the internal and reserved fields, is the document ID number. A *document ID* is a unique number that identifies each document. Notes creates this document ID internally, and it cannot be altered. The document ID number is used within Notes to link documents to their parent and children documents, as well as to return field values from specific documents. There are several @Functions listed in Table 8.6 that work specifically with the document ID. An example of a document ID is shown in Figure 8.4.

Table 8.6 Document ID @Functions

@Function and Syntax	Notes
@DocumentUniqueID	This function returns the 32-character document UNID, which is part of the document's complete ID. To see the document's ID number, open the Document Properties InfoBox and click the Info tab. The UNID is the first line of the ID number, as shown in Figure 8.4. When you use this function in a formula, you can create a link to the document specified by the ID number, within the current database.
@InheritedDocumentUniqueID	This function returns the UNID for the parent of the current document. This function is used within response documents to define and maintain the link to the parent. To make this function work properly, inheritance must be enabled using the Inherit Values from Selected Document option on the Form Properties InfoBox.
@GetDocField (documentUNID; fieldname)	This function returns the value of the field specified in the function from the document specified by the document UNID.
@SetDocField (documentUNID; fieldname; new value)	This function sets the value of the field specified in the function on the document specified by the document UNID.

NOTE

Enclose the fieldname used in the last two functions in quotation marks, unless it is stored in a variable.

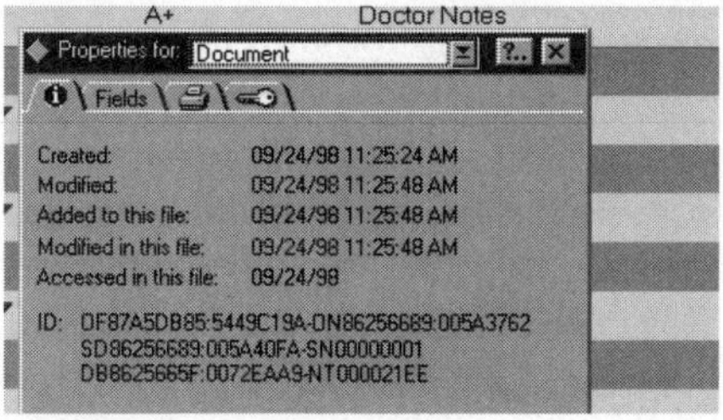

Figure 8.4 Document ID.

List @Functions

Some of the advanced @Functions you need to know to develop advanced applications are those that allow you to manipulate lists. A *list* is any series of elements of the same data type that has a name. Lists can be acted upon as a single object by an @Function or by other operator. As discussed in Chapter 3, *Fields, Formulas, Agents, and Actions*, the colon (:) is the list operator that links the list items. An example of a text string list would be "Author" : "Instructor" : "Programmer". Some of the @Functions that relate to list manipulation were also used in Application Development I. We will review those, therefore, and add some others for the Application Development II exam. Table 8.7 describes some of the list-manipulation @Functions.

Table 8.7 List @Functions

@Function and Syntax	Notes
@Contains (string; substring)	This function searches for a substring within a specific text string, and returns a true or false value. Both arguments in the function can be either text or text lists. Note that this function is case-sensitive. The following example would return a value of one: `@Contains ("Libby Nicole Schwarz"; "N")` Often, the @Contains is used as part of a longer @If function, as in the example: `@If (@Contains (User; "Libby"); "This is the author's opinion."; "This is not the author's opinion.")` This function would examine the User field for the name Libby to determine which response to print.

continues

Table 8.7 Continued.

@Function and Syntax	Notes
@Elements(list)	This function returns a numerical value for the number of elements in the list. In the following function, for example, the returned value would be four: `@Elements ("Mike" : "Brian" : "Libby" : "Scott").`
@Explode (string; separator; includeEmpties) or @Explode (daterange)	This function returns either the individual words or clauses of the string if using a text string, or individual days if using a date range. For example, the function @Explode (Authors) returns Libby; Scott; Brian; Mike; if the contents of the Authors field are "Libby Scott Brian Mike". Note that the default display separator is a semicolon (;).
@Implode (textlist; separator)	This function works in the opposite manner of @Explode and returns a concatenated text string from a list.
@IsMember (textlist; textlist) OR @IsNotMember (textlist; textlist)	These functions, which are case sensitive, return a true or false value to indicate whether a piece of text or a text list is or is not contained within another text list. In the following example, the function returns a "1" if the word Author is in the JobDescription field: `@IsMember("Author"; JobDescription).` To make sure that the words in the JobDescription field match the case shown in the text string, you might use the following formula: `@IsMember("Author"; @ProperCase (JobDescription)).`

@Function and Syntax	Notes
@Keywords(textlist; textlist; separator)	This function returns the items from the second argument that are also found in the first argument. In the following formula, the return value would be Lotus, Microsoft, NT: `@Keywords ("Lotus Notes 5.0 will be released next year" : "Windows NT 5.0 will be released next year" : "Microsoft Exchange competes with Lotus Notes"; "Lotus"; "Novell"; "Microsoft"; "IBM"; "NT")`
@Member (value; list)	This function returns a numerical value relating to the position of the value in the list. The function returns a value of three if the name Libby is the third item in the field SeniorityList: `@Member ("Libby"; SeniorityList)`
@Replace (list; replacewhat; replacewith)	This function replaces the items in the first argument (list) that are also in the second argument (replacewhat) with the specified items in the third argument (replacewith). The items in the lists are used in order, so the first item in the *replacewhat* list is replaced by the first item in the *replacewith* list.
@Subset (list; number)	This function moves from left to right through the list and returns the number of items specified. If the number is a negative number, the function searches from right to left. The following function returns Brian; Scott : `@Subset ("Bob" : "Ron" : "Joe" : "Brian" : "Scott"; -2)`
@Unique (textlist)	The argument in this function is optional. If used without the argument, @Unique returns a random, unique text value. When used with the textlist, the function returns the first occurrence of each item in the list.

Environment Variable @Functions

Environment variables give the designer the ability to use information among multiple Notes databases or Notes sessions. Environment variables are pieces of data stored in a user's NOTES.INI file. The NOTES.INI file stores a variety of data about the user and the choices made for that installation of Notes. The environment variables use this permanent storage location for holding information that will be reused or passed between databases. Figure 8.5 shows a NOTES.INI file with environment variables. Note that the environment variables can be distinguished from the other information stored in the NOTES.INI file by the $ that begins the name.

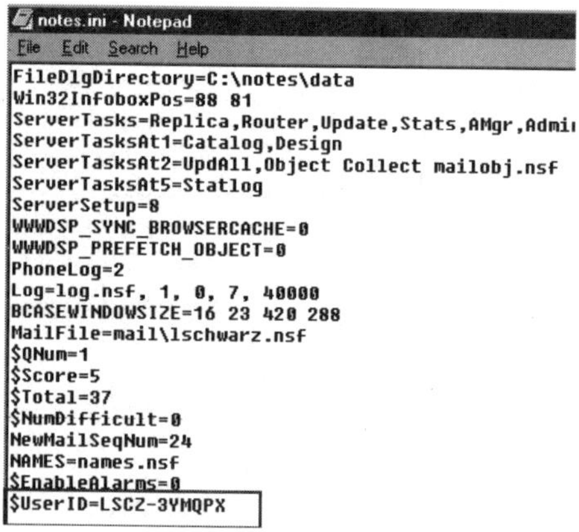

Figure 8.5 NOTES.INI file showing environment variables.

NOTE

Environment variables are stored with dollar signs in their names in the NOTES.INI file. When you refer to them in functions, however, the dollar signs are not necessary.

The following are some of the main uses of environment variables in Notes:

- Passing data between or among different formulas or Notes databases.
- Generating a series of sequential numbers for a user.
- Reusing information that a user might otherwise have to reenter.

Environment variables in Notes can be set in the following three ways:

- Using the ENVIRONMENT keyword to set an environment variable by placing it at the beginning of an expression. Use the following syntax:

```
ENVIRONMENT variable := value
```

The following expression, for example, takes the value of a field called SocNumber, in which the user typed his or her social security number, and stores it in the environment variable EmployeeNumber:

```
Environment EmployeeNumber := @Text (SocNumber)
```

- Using the function @SetEnvironment within another @Function, such as @If, to set an environment variable. Note that the @SetEnvironment function cannot be used in column or selection formulas. Use the following syntax:

```
@SetEnvironment (VariableName; value)
```

The following example performs the same function as the earlier expression:

```
@SetEnvironment ("EmployeeNumber"; @Text(SocNumber))
```

- Use the function @Environment to set an environment variable in almost any formula. Use the following syntax:

```
@Environment (VariableName; value)
```

The following function serves the same purpose as the earlier expressions:

```
@Environment ("EmployeeNumber"; @Text(SocNumber))
```

After setting an environment variable, use the @Environment function with only one argument to retrieve the variable's value for use. The @Environment function will use the following syntax when retrieving values: @Environment (variable). To retrieve the environment variable set in the earlier examples (EmployeeNumber), use the following function:

```
@Environment ("EmployeeNumber")
```

NOTE

Be aware of where you are using an environment variable. Environment variables cannot be set in column or selection formulas. In addition, do not try to set them in certain formulas that access a server's NOTES.INI file, such as the following:

- Replication formulas.
- Agents triggered automatically by new mail arriving or by a schedule.
- Selection formulas.
- Column formulas.

@Db Functions

The functions in the previous section enabled the designer to store information to be reused in other Notes sessions or databases. The environment variables are most useful when the information being accessed or stored is user-specific. When a designer needs to pull information from other databases that is not user-specific, the designer may prefer to use @Db functions. The @Db functions, such as @DbColumn, @DbLookup, and @DbCommand, can be used to pull information from current Notes databases, other Notes databases, and, in most cases, non-Notes databases. @Db functions can be used in most formulas, with the exceptions of mail agents, view selection formulas, and column formulas.

The @Db functions covered on the Application Development II exam are @DbColumn and @DbLookup. The other @Db functions, listed next, are not tested specifically on the Application Development II exam. It is important, however, to recognize them.

- @DbCommand returns the result of a search made on a non-Notes database.
- @DbExists determines whether a database exists when given a server and filename.
- @DbManager returns the list of users, groups, and servers that have Manager access to a particular database.
- @DbName returns the name of the current Notes server and database.
- @DbTitle returns the title of the current Notes database.

@DBCOLUMN

The @DbColumn function returns all the values in a column of a view, folder, or table, in the case of a non-Notes database. @DbColumn is often used for populating keyword lists automatically and dynamically. Use the following syntax for this function:

```
@DbColumn (class : "NoCache"; server :database; view;
ColumnNumber)
```

- Class defines the type of database being accessed by the function —either Notes or ODBC. For a Notes database, you can also use "".
- The *NoCache* keyword, which is optional, forces the results of the lookup to be newly retrieved each time. If the keyword is not used, the results of the function will be cached and reused. Use this keyword based on the performance you need. If you retrieve data from a database that rarely changes and is fairly large, for example, you would probably want to allow the value to be cached to save lookup time. If the data changes often, however, you will need to use this keyword to ensure that the most recent information is presented to the users.
- The server parameter is either "" for a local database or the full path to the server for a server database.
- The database parameter is the full path to the database (if you are not referring to the current database). You may also use the replica ID or the @DbName function in this parameter.
- The view parameter requires the name of the view that contains the column from which you want to retrieve data. Use the full name of the view as shown in the View Properties InfoBox, including cascading names separated by double backslashes (\\). If the view has a synonym, use this option instead.
- The ColumnNumber parameter requires the number of the column containing the data you want to retrieve. Column numbers are counted from left to right, starting at one. Certain columns will be excluded from this count, including hidden columns, columns that display a constant value, or columns that use any of the following @Functions: @DocChildren, @DocDescendants, @DocLevel, @DocNumber, @DocParentNumber, @DocSiblings, @IsCategory, or @IsExpandable. If you use a column number indicating a column that does not exist, you will get a blank, or null, value—not an error.

One of the most common uses for the @DbColumn function is to provide dynamic keywords for a field. If you want to display a list of users or documents based on data in another view or database, the @DbColumn function is the one you need to use. To use a formula to determine keyword choices, click the choices drop-down list on the Keyword options tab of the Field Properties InfoBox and choose Use Formula for Choices.

If the database that contains the software list is in a database called SOFTWARE.NSF on the current server, for example, the following formula in the keyword field would enable users to choose software:

```
@DbColumn ("Notes" : "NoCache"; "" : "Software.NSF"; "By
Product";2)
```

This function returns as keywords the values contained in the second column of the By Product view. Additionally, you may want to nest the @DbColumn function into @If or other functions. This option can provide a choice of keywords based on the value of another field, for example.

@DBLOOKUP

The @DbLookup formula returns the value of a particular field or column in a view based on a unique, specified key. The formula can use either of the two examples of syntax:

```
@DbLookup ( class : "NoCache"; server :database; view; key;
FieldName) OR
@DbLookup ( class : "NoCache"; server :database; view; key;
ColumnNumber)
```

- Class defines the type of database being accessed by the function —either Notes or ODBC. For a Notes database, you can also use "".
- The *NoCache* keyword, which is optional, forces the results of the lookup to be newly retrieved each time. If the keyword is not used, the results of the function will be cached and re-used. Use this keyword based on the performance you need. If you are retrieving data from a database that rarely changes and is fairly large, for example, you probably want to allow the value to be cached to save lookup time. If, however, the data changes often, you will need to use this keyword to ensure that the most recent information is presented to users.
- The server parameter is either "" for a local database or the full path to the server for a server database.
- The database parameter is the full path to the database, if you are not referring to the current database. You may also use the replica ID or the @DbName function in this parameter.

- The view parameter requires the name of the view that contains the column whose data you want to retrieve. Use the full name of the view (as it appears in the View Properties InfoBox), and separate cascading names with double backslashes (\\). If the view has a synonym, use this instead.
- The key parameter is the parameter that determines which document is read when retrieving the value. The key for a document is the value to match in the first sorted column in the view. Usually, you will specify a field name for the key (distinguish the fact that it is a field name by leaving off the quotation marks). When you specify a field name for the key, Notes looks up the value of that field in the current document. It then matches that value to the values in the first sorted column in the specified view.
- The FieldName parameter specifies what data is returned from the matched document(s). After documents have been selected based on matching the key value, Notes can return the data from any field in that document. Keep in mind that lookups using columns may be faster, as Notes does not have to search the entire document for the specified field. In addition, the specified field may not be available if the document was created using a different form.
- The ColumnNumber parameter requires the number of the column containing the data you want to retrieve. If the lookup value is *Software* and the ColumnNumber is 2, the formula would return all data contained in the second column for documents that are categorized under the word software.

  Column numbers are counted from left to right, starting at one. Certain columns will be excluded from this count, including hidden columns, columns that display a constant value, or columns that use any of the following @Functions: @DocChildren, @DocDescendants, @DocLevel, @DocNumber, @DocParentNumber, @DocSiblings, @IsCategory, or @IsExpandable. If you use a column number indicating a column that does not exist, you will get a blank, or null, value—not an error.

@DbLookup functions are often used to complete fields in a document based on one piece of data entered. If you want to complete the phone number, e-mail address, and office location of the User Question, for example, you could use the Last name as a key to look up these values in the *Public Name and Address Book* (NAB). The formula might look something like this:

```
@DbLookup ("Notes" : "NoCache"; "" : "names.nsf"; "People";
UserName; 2)
```

This example returns the values of the second column in the People view of the NAB, based on the value of the UserName field, which is the key.

The following list describes some of the issues to be aware of when using any @Db functions:

- Use a replica ID number for the server name if there may be multiple copies of the database on the servers.
- Be warned that @DbColumn functions can only return 64KB of data. This limitation means, for example, that trying to return data from a view from the Help database or the Names.nsf database would probably not be successful. The Notes Help database provides a formula for determining how much of your data will be displayed using an @DbColumn function to return text.

  ```
  2+ (2*number of entries returned) +total text size of all
  entries
  ```

- To use a slash in the database name or view name in this function, use a double backslash (\\). For example, for the database "Accounting\Orders.NSF," use "Accounting\\Orders.NSF".
- Server database security may affect the lookup function. If the database that contains the @DbColumn function is the source database, and the database containing the view and data is the target database, the source database must have at least Reader access to the target database. This goal can be accomplished by giving the target database's—Default—group at least reader access, or by giving the source database specific access using the replica ID as the name in the Access Control List.

@PickList

The @PickList function displays a dialog box to enable a user to choose a document from either a view specified by the designer or from the Address dialog box. Use the following syntax for an @PickList function:

```
@PickList ([NAME] : [SINGLE]) or
@PickList ([CUSTOM]; server : database; view; title; prompt;
column)
```

These parameters are described next.

- The [NAME] keyword tells Notes to display the Names dialog box from the NABs available, as shown in Figure 8.6. The parameter displays the double-sided dialog box that enables the user to choose more than one name.

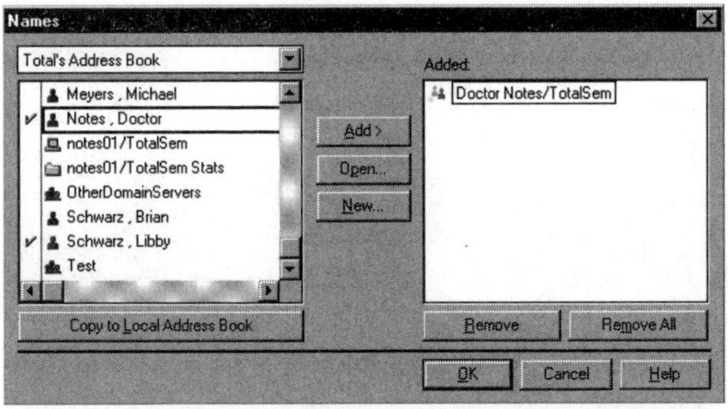

Figure 8.6 Names dialog box allowing multiple sections.

- The [SINGLE] keyword, which is optional, tells Notes to display a Names dialog box that only enables the user to choose one name from the Address book. This dialog box is shown in Figure 8.7.

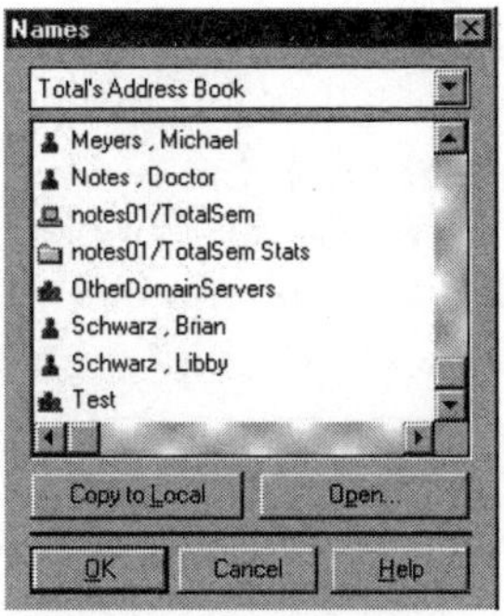

Figure 8.7 Names dialog box enabling only one selection.

- The [CUSTOM] keyword tells Notes to display a view rather than the NAB. The dialog box displayed by this type of @PickList function is shown in Figure 8.8.

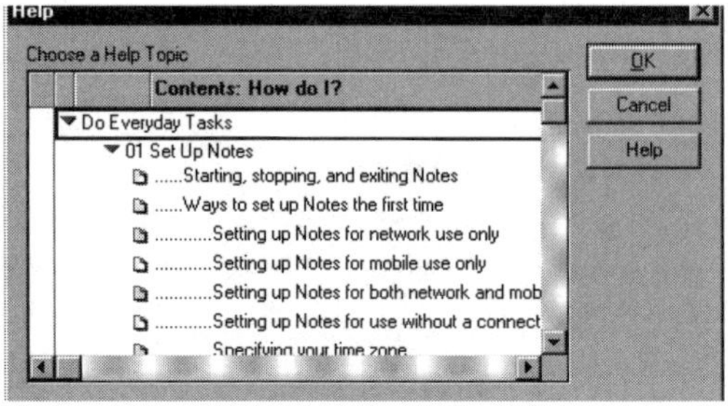

Figure 8.8 @PickList ([CUSTOM]) dialog box.

The formula used to create the @PickList dialog box displayed in Figure 8.8 is shown in Figure 8.9.

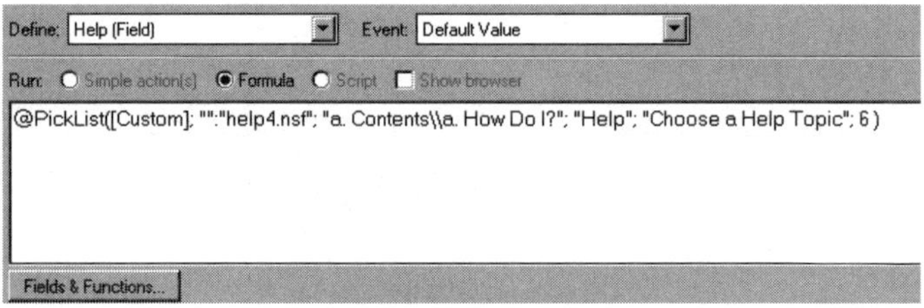

Figure 8.9 @PickList ([CUSTOM]) formula.

- The server parameter is either "" for a local database or the full path to the server for a server database.
- The database parameter is the full path to the database, if you are not referring to the current database. Unlike the @Db functions, you may *not* use the replica ID in this parameter.
- The view parameter requires the name of the view that contains the column whose data you want to retrieve. Use the full name of the view as shown in the View Properties InfoBox, including cascading names separated by double backslashes (\\). If the view has a synonym, use this instead.
- The title parameter is the window title that Notes will display at the top of the dialog box. This field should be a descriptive text title, enclosed in quotes.
- The prompt parameter will appear at the top of the dialog box to explain the choice to the user. This field should be text, enclosed in quotes.

- The column parameter is the column number, counted from left to right and including all columns, including hidden columns. Unlike @Db functions, all columns, regardless of value or type of formula, should be counted to determine the ColumnNumber. The value in the specified column number will be returned as the value in the field that contains the @PickList function.

An example of the use of @PickList follows:

```
@PickList ([CUSTOM]; “”; “By Software”; “Software Selection”;
“Please select the appropriate topic below:”; 2”)
```

This function looks in the current database, in the By Software view and returns the value of the second column for the documents chosen by the user. The window title would be *Software Selection* and the prompt would be *Please select the appropriate topic below:*.

NOTE
The @PickList function can be used in buttons, manual agents, and form or view actions. The function, however, cannot be used in column or selection formulas, mail or scheduled agents, Hide-when formulas, form formulas, or window title formulas.

The @PickList function is very similar to the @DbLookup and the @DbColumn functions. There are several differences, however, which can make @PickList a better choice in some situations:

- The @Db functions can only return up to 64K of data, whereas the @PickList does not have any similar limitation.
- @PickList performs the lookup each time it is used, as there is no *NoCache* option; however, it performs a lookup faster than an @Db function.
- @PickList has a QuickSearch capability that enables users to type in a few letters to search for a document.

@Prompt

The @PickList function described earlier provides one method of enabling the application to interact with the user to obtain information. Another @Function that provides this capability is the @Prompt function. The @Prompt function displays a dialog box that enables the user to enter a value. It will return either a yes/no type of value or text, depending on the style chosen by the designer. Use the following syntax for an @Prompt function:

```
@Prompt ( [style] : [NoSort]; title; prompt; DefaultChoice;
ChoiceList)
```

- The designer can choose one of nine styles for the [style] parameter. There are nine styles available for use with the @Prompt function. The other parameters for the function vary based on the style that the designer chooses. The styles available are: OK, YesNo, YesNoCancel, OkCancelEdit, OkCancelList, OkCancelCombo, OkCancelEditCombo, OkCancelListMult, and Password.
- The [NoSort] keyword, which is optional, can be used any time choices are provided to the user. This keyword forces Notes to display the choices as entered. If you do not put the keyword in, the choices will be sorted alphabetically.
- The title parameter provides a window title for the @Prompt dialog box that the function displays to the user. This parameter is required.
- The prompt parameter provides text to be displayed in the dialog box, such as instructions or comment. This parameter is required.
- The DefaultChoice parameter provides a default value for the user to change or accept. It is not used with the following styles: OK, YesNo, YesNoCancel, and Password, but is required for all other styles.
- The ChoiceList parameter provides a list of choices for the user with the following styles of @Prompt: OkCancelList, OkCancelCombo, OkCancelEditCombo, and OkCancelListMult. The values of this parameter can either be text strings or @Functions that return text strings, separated by colons.

The styles of @Prompt to use vary with the type of information required from the user. Similarly, the parameters to be used vary with the style chosen. Each style is described in the following paragraphs:

- OK: This style provides information to the user. The only value returned is an acknowledgment that the information has been displayed. The parameters to be used with the style are title and prompt.

  The OK style can be used to give the user a reminder or confirmation. You may use the OK style to remind the user to replicate the databases after creating new documents, as shown in Figure 8.10, using the following @Prompt:

```
@Prompt ( [OK]; "Reminder"; "You will want to run a replication
on this database after creating new documents.")
```

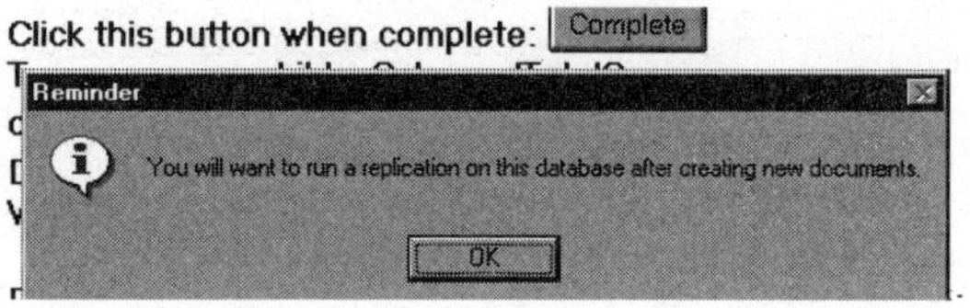

Figure 8.10 @Prompt ([OK]).

- YesNo: The YesNo style returns a yes (1) or no (0) value after providing the user with information. The YesNo style requires a title and a prompt parameter. One use of the YesNo style might be to confirm that a user wants to perform a particular document. In the case of running an agent to approve a sale, for example, you might create the following @Prompt function, as displayed in Figure 8.11:

  ```
  @Prompt ([YesNo]; "Approve"; "Do you really want to approve
  this sale?")
  ```

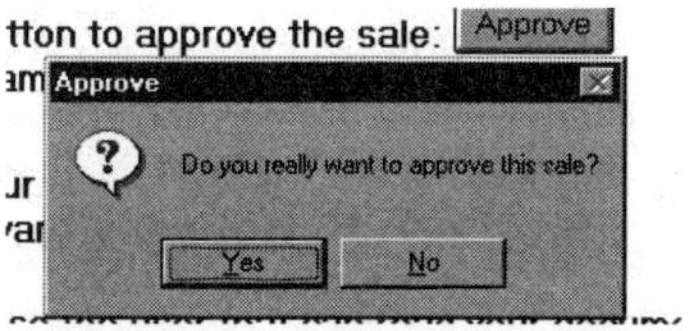

Figure 8.11 @Prompt ([YesNo]).

- YesNoCancel: The YesNoCancel style is very similar to the YesNo style in appearance and return value. The Cancel button returns a –1 value, while the Yes (1) and No (0) buttons return the same values as explained earlier. The YesNoCancel style requires the title and prompt parameters. You may use a YesNoCancel in a similar manner as YesNo to provide information to the user, as shown in Figure 8.12:

  ```
  @Prompt ([YesNoCancel]; "Approve"; "Do you really want to
  approve this sale?")
  ```

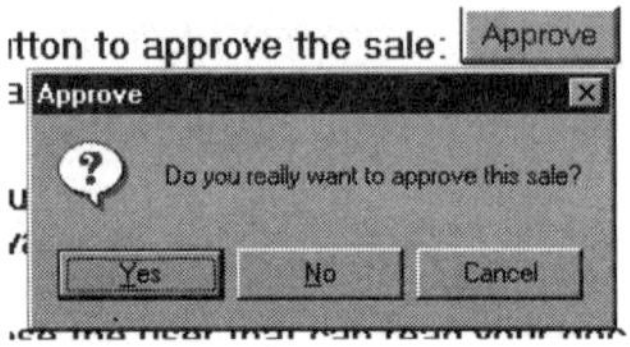

Figure 8.12 @Prompt ([YesNoCancel]).

- OkCancelEdit: The OkCancelEdit style returns a text string entered by the user. The title and prompt parameters are required, and the DefaultChoice parameter can be included to provide a default entry for the user. Note that if you choose not to provide a default entry, you must provide a null value for this parameter. To obtain the user's favorite color, for example, you might use the following @Prompt function, as displayed in Figure 8.13:

```
@Prompt ([OkCancelEdit]; "Color Request"; "Please type in your
favorite color:"; "")
```

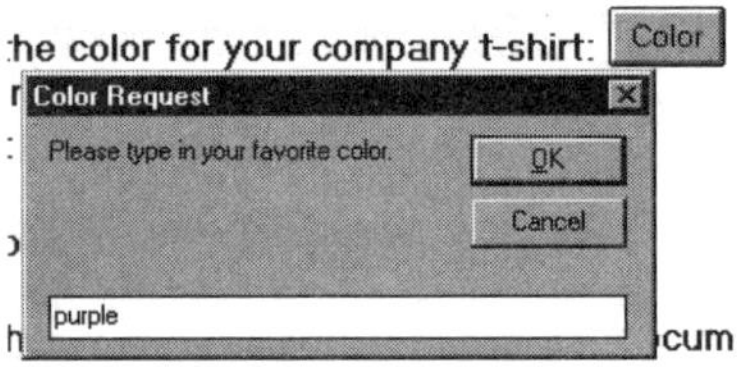

Figure 8.13 @Prompt ([OkCancelEdit]).

- OkCancelList: The OkCancelList style returns a text string based on providing the user a list of choices in a list box style. This style requires the ChoiceList parameter to be included to provide the list of choices. To obtain the user's favorite color using this style, for example, you might use the following @Prompt function, as displayed in Figure 8.14:

```
@Prompt ([OkCancelList]; "Color Request"; "Please select your
favorite color:"; "Green"; "Yellow" :"Green" : "Blue" : "Purple")
```

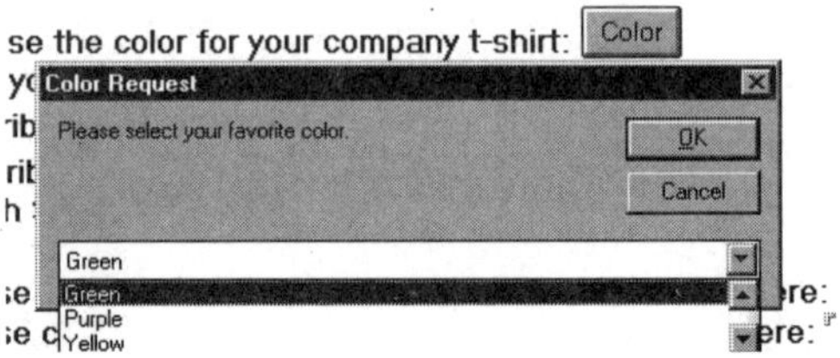

Figure 8.14 @Prompt ([OkCancelList]).

- OkCancelCombo: The OkCancelCombo style returns a text string based on providing the user a list of choices in a combo box style. The default value is the only initial value displayed to the user. To obtain the user's favorite color using this style, for example, you might use the following @Prompt function, as displayed in Figure 8.15:

```
@Prompt ([OkCancelCombo]; "Color Request"; "Please select your
favorite color:"; "Green"; "Yellow" :"Green" : "Blue" : "Purple")
```

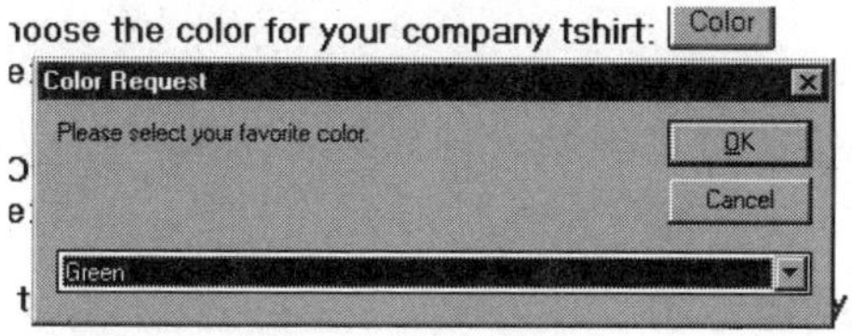

Figure 8.15 @Prompt ([OkCancelCombo]).

- OkCancelEditCombo: This style provides the same function and appearance as the OkCancelCombo, except the users have the ability to enter a value that is not in the list. To obtain the user's favorite color using this style, for example, you might use the following @Prompt function, as displayed in Figure 8.16:

```
@Prompt ([OkCancelEditCombo]; "Color Request"; "Please select
your favorite color from the list or type your own:"; "Green";
"Yellow" :"Green" : "Blue" : "Purple")
```

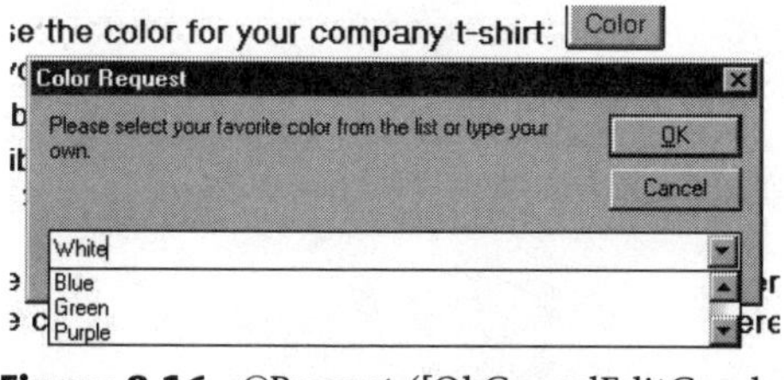

Figure 8.16 @Prompt ([OkCancelEditCombo]).

- OkCancelListMult: This style returns one or more text string values. The style provides a list, similar to the one from the OkCancelList style, but the user is given the ability to choose more than one option. To obtain the user's favorite color using this style, for example, you might use the following @Prompt function, as displayed in Figure 8.17:

```
@Prompt ([OkCancelListMult]; "Color Request"; "Please select
your favorite color or colors from the list:"; "Green";
"Yellow" :"Green" : "Blue" : "Purple")
```

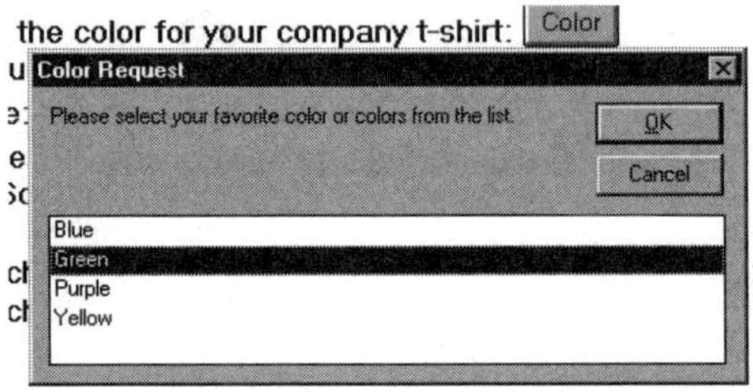

Figure 8.17 @Prompt ([OkCancelListMult]).

- Password: The password style enables the designer to request a user to enter a password. The password will not be displayed on the screen but will be captured by the function. Do not confuse this password with the password used with the user's ID file or with the password used by the @Password function. To allow a user to enter a password for a particular database or application, you might use a function like the one listed next. Note that the warning, as shown in Figure 8.18, is provided automatically by Notes.

```
@Prompt ([Password]; "Password Request"; "Please type your
password for this database:")
```

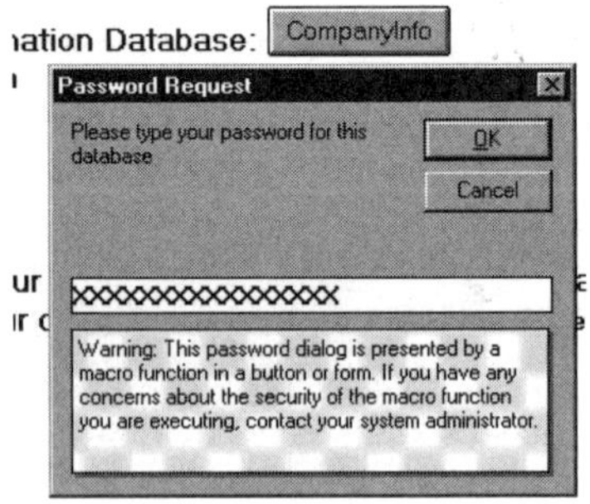

Figure 8.18 @Prompt ([Password]).

The @Prompt function is one of the functions that enables the designer to request input from the user at certain times. The @Prompt function can be used in field formulas, buttons, SmartIcons, manual agents, form actions, and view actions. This function does not work in column or selection formulas, mail or scheduled agent formulas, and has limited usefulness in window title and form formulas. A good example of use might be to confirm an action that a manual agent or action will perform.

@DialogBox

Another function that enables the designer to present the user with a dialog box intended to gather information is the @DialogBox function. This function displays the current document using a different form. The function can be used to display a form, subform, or a layout region. If the underlying document is in edit mode when the dialog box is displayed, any information entered into the dialog box will then be stored with the document. Use the following syntax with @DialogBox:

```
@DialogBox ("form"; [AutoVertFit] : [AutoHorzFit] : [NoCancel]
: [NoNewFields] : [NoFieldUpdate] : [ReadOnly] ; "title")
```

- The form parameter enables you to name the form that you would like to display to the user in the dialog box. If the form or subform indicated in this parameter has a layout region on it, the layout region will be displayed. If the form or subform has more than one layout region, only the first one will be displayed to the user.
- The AutoVertFit and AutoHorzFit optional keywords size the dialog box to the size of the form or layout region automatically. It is recommended to use both of these keywords together.
- The NoCancel optional keyword removes the Cancel button from the dialog box. By default, the dialog box is displayed to the user with both an OK and a Cancel button. These buttons are defined and inserted automatically by Notes.
- The NoNewFields optional keyword does not allow any fields in the dialog box that are not in the underlying form to be added to the form.
- The NoFieldUpdate optional keyword does not allow any edits made in the dialog box to be passed on to the underlying form.
- The ReadOnly optional keyword forces the dialog box not to accept edits. By default, the dialog box is read-write enabled.
- The title parameter enables the designer to give the dialog box a window title. Without this parameter, the title of the dialog box defaults to Lotus Notes.

One use for the @DialogBox function, for example, might be to present Help screens to the user to assist with completing a form. To do this, you might use a function like the one listed next, as shown in Figure 8.19:

```
@DialogBox( "Help"; [AutoVertFit] : [AutoHorzFit] : [NoCancel]
: [NoNewFields] : [NoFieldUpdate] : [ReadOnly]; "Help Screen")
```

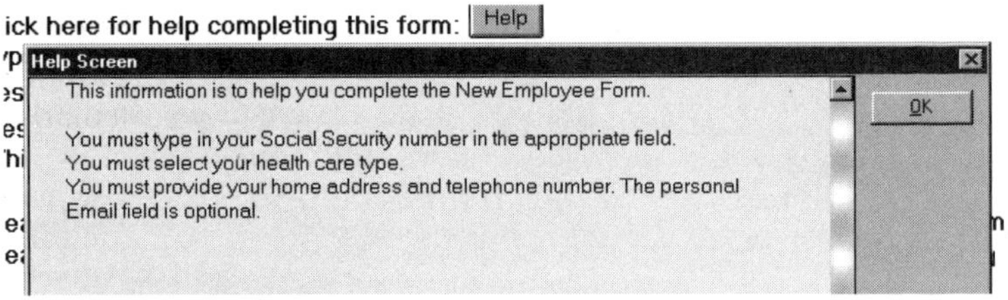

Figure 8.19 @DialogBox example.

Calculation Control @Functions

As you begin to write these more complex functions and formulas, you will need some tools to make the functions act in the desired manner. These are considered calculation control functions. Some functions that can be used as calculation control functions are: @Return, @IsNewDoc, @IsDocBeingSaved, @IsDocBeingMailed, @IsDocBeingLoaded, @IsDocBeingRecalculated. Each of these functions checks the status of the document or a field and either stops or continues based on this status. For the most part, these functions will be used in conjunction with the @If function that we covered for the Application Development I exam. To review, the @If function examines a condition and returns one value if it is true and a different value if it is false. The @If statements can be nested up to 99 times but must always have an odd number of arguments. The following is an example of an @If statement used to decide the window title for a form:

```
@If( @IsNewDoc; "New Question Form"; Question)
```

This formula checks to see whether the document is new. If it is new, it displays the title as *New Question Form*, otherwise it displays the text of the Question field as the title.

In addition, these functions are also often part of multi-line functions. Two things are especially important to know when using multi-line functions. First, separate each line of a multi-line function with a semicolon. Sometimes a single function or expression will wrap to the next line in the formula dialog box; this situation does not require a semicolon. In cases where you intentionally place multiple expressions or functions in a single formula, however, insert a semicolon at the end of each line. Second, the value that is returned in the case of a multi-line expression is the value of the last line of the formula. The preceding lines are used to evaluate the last line, but the last line must always be the main expression of the formula.

The following paragraphs describe the details of these calculation control functions:

- @Return enables the designer to stop the evaluation of the function at a certain point. This feature is generally used to check a value of a field in the document and then use that value to determine whether the function should continue or stop. For example:

  ```
  @If (CallStatus = "Complete"; @Return(""); "You must complete
  this call before forwarding the message.")
  ```

  This function checks the value of the field CallStatus. If it is complete, the function stops. If it is not complete, the function returns the specified text string. You might include this function in buttons or field formulas, agents that run formulas, and SmartIcons.

- @IsNewDoc enables the designer to check whether the document has been saved or not. It will return a true (1) value if the document is being edited and has not been saved, or a false (0) value in any other case. The most common use is when determining an appropriate window title for a form that is being composed and a different window title when that form is read. This function can be used to evaluate the current status of the document when it is used in SmartIcons, button, Hide-When, section editor, window title, field, form, and form action formulas. In column, selection, agent, and view action formulas, it *always* returns a false (0) value.

- @IsDocBeingSaved enables the designer to check whether the document is being saved. It will return a true (1) value if the document is being saved and a false (0) value any other time. This function is most useful when a designer wants to track how often a document has been saved by incrementing a value each time the document is saved.

- @IsDocBeingMailed is a new function to R4 that enables a designer to check whether a document is being mailed. It will return a true (1) value if the document is being mailed and a false (0) value any other time. This feature might be useful if a designer wants to track how often a document has been mailed or forwarded.
- @IsDocBeingLoaded enables the designer to check whether a document is in the process of being loaded into memory. It will return a true (1) value if the document is being opened and a false (0) value any other time. This feature is useful for tracking how many times a document has been opened.
- @IsDocBeingRecalculated enables the designer to check whether a document is being refreshed or recalculated. It will return a true (1) value if the document is being refreshed and a false (0) value any other time. This feature enables the designer to track the number of times a document has been refreshed or enables the designer to prevent a field from being updated when the document is refreshed.

Advanced Agents and Actions

The previous sections of this chapter described fields and functions that allow you to automate Notes applications or add other functionality. In addition to the fields and functions we discussed earlier, you can use agents and actions to automate your applications. In Chapter 3, *Fields, Formulas, Agents, and Actions,* we discussed the fact that an agent is defined by three pieces of information: a trigger that causes the agent to act, the documents that are selected for the agent to act upon, and what process the agent performs. For the Application Development I exam, we created simple agents that were triggered manually. In this section, we will review how to create these agents and then build on that to trigger the agents automatically. We will also cover some methods for troubleshooting the agents that you create. In addition, you learned that an action can be inserted into a view or a form and can also be used to automate a function. In many cases, actions will be used to run agents. This section will also describe creating actions and using them to run agents.

Creating and Using Agents

Agents can be used to automate a variety of tasks. Some examples of good uses for agents might include the following:

1. Automating repeatable processes.
2. Automating processes that are too complex for the average user.
3. Automating processes that occur on schedule or after a repeated event.

To create an agent, choose Create . . . Agent from the menus. This feature displays the Agent design window, as shown in Figure 8.20. In the agent design window, choose a name for the agent that will be descriptive for the user if the agent will run manually. You must also determine whether the agent will be shared or private. This choice can *only* be made when the agent is created. After you save an agent, this check box is grayed and cannot be changed.

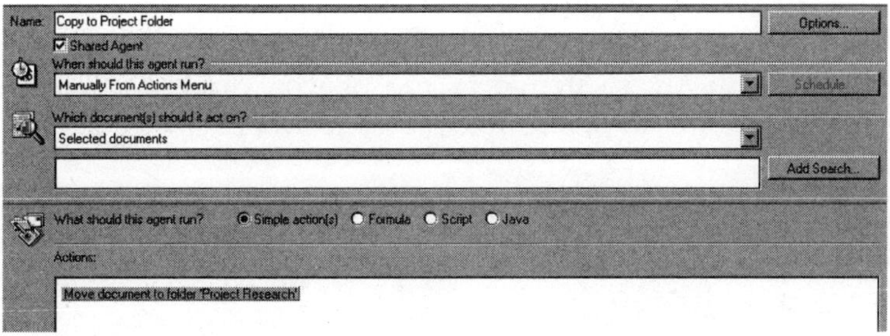

Figure 8.20 Agent design window.

You can also set other properties of the agent by clicking the Options button. This displays the Options dialog box, as shown in Figure 8.21. In this dialog box, you can make the following choices:

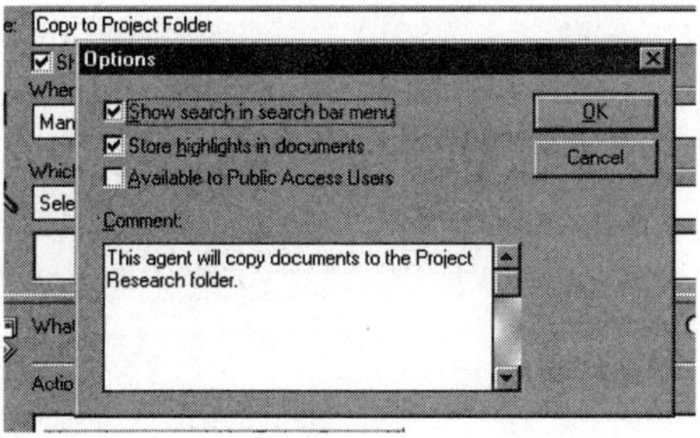

Figure 8.21 Agent options dialog box.

- *Show search in search bar menu.* This feature enables a user to view the agent's search formula when they use the search bar in this database. This feature would enable a user to preview the documents on which the agent would act.
- *Store highlights in documents.* This option enables the user to see a red highlight around the search words in each document after the agent has run.
- *Available to Public Access Users* (V4.5 and higher). This option enables the agent to be available for users that only have public access to the database. It will not be tested on the exam.
- *Comment.* This field enables the designer to add a description of the agent's purpose and function. The first few words of the description can be viewed in the Agents view.

The next section of the agent design window is used to determine when the agent should be run or what should trigger the agent. The choices available in this drop-down list are as follows:

- Manually from Actions list
- Manually from Agents list
- If New Mail has arrived
- If Documents have been created or modified
- If Documents have been pasted
- On Schedule Hourly, Daily, Weekly, or Monthly
- On Schedule Never

The first two choices are manual triggers. That means that a user has to activate the agent by clicking a button or selecting the agent from the Actions menu. The other choices are triggered either by an event or on a schedule. When a application event occurs—such as when mail is received or documents are saved, created, or pasted—the agents that use the If New Mail has arrived, If Documents have been created or modified, and If Documents have been pasted triggers will activate. The other selections activate on schedule, based on days, weeks, hours, or months. If you choose to activate an agent based on a schedule, in addition to choosing On Schedule Hourly, Daily, Weekly, Monthly, or Never, you must click the Schedule button to create the schedule.

If you choose to schedule an agent on an hourly basis, the hourly options dialog box, shown in Figure 8.22, is displayed. The hourly agent can be run every 30 minutes, one hour, two hours, four hours, or eight hours. In addition, you can choose to have the agent run only during a certain period or only on weekdays. The dialog box also enables you to specify the server on which the agent should run. By default, this field contains the current server. If the agent will need to be run on a different server, this field must be updated. If you select the Choose when agent is enabled checkbox, you can select the server when the agent is enabled.

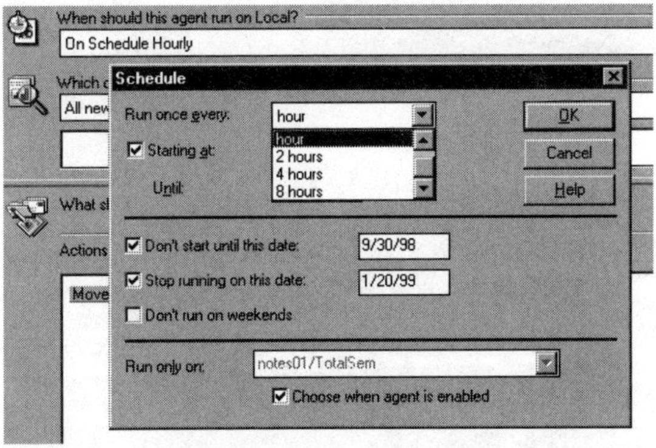

Figure 8.22 Hourly agent options.

If you choose to schedule an agent on a daily basis, the daily options dialog box, shown in Figure 8.23, is displayed. You can specify the time to run a daily agent in this dialog box. In addition, you can choose to have the agent run only during a certain period of days, or only on weekdays. The dialog box also enables you to specify the server on which the agent should run. By default, this field contains the current server. If the agent will need to be run on a different server, this field will need to be updated. If you select the Choose when agent is enabled checkbox, you can select the server when the agent is enabled.

If you choose to schedule an agent on a weekly basis, the weekly options dialog box, shown in Figure 8.24, is displayed. You can specify the day and time to run a weekly agent in this dialog box. In addition, you can choose to have the agent run only during a certain period of days. The dialog box also enables you to specify the server on which the agent should run. By default, this field contains the current server. If the agent will need to be run on a different server, this field must be updated. If you select the Choose when agent is enabled checkbox, you can select the server when the agent is enabled.

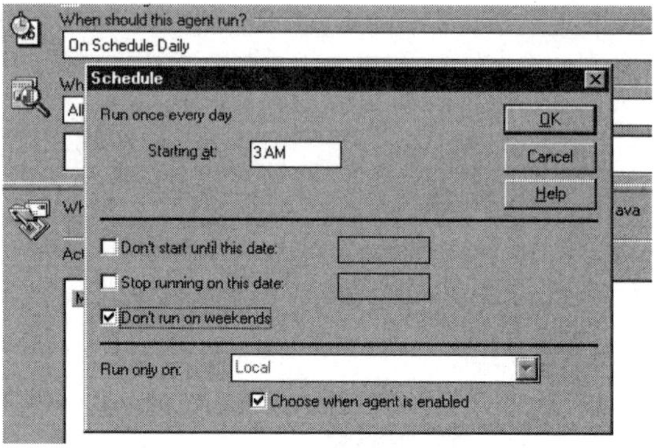

Figure 8.23 Daily agent options.

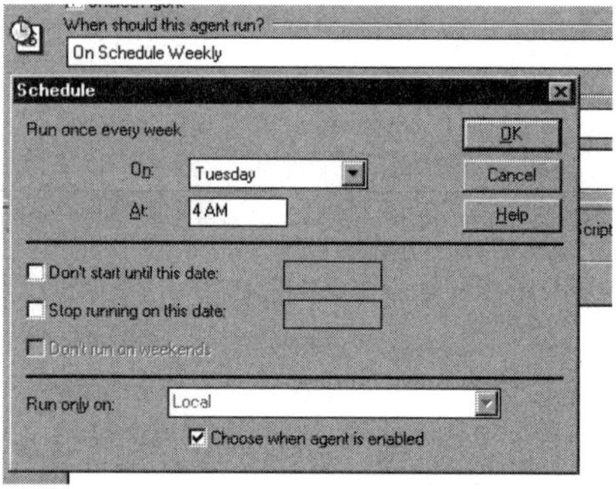

Figure 8.24 Weekly agent options.

If you choose to schedule an agent on a monthly basis, the monthly options dialog box, shown in Figure 8.25, is displayed. You can specify the day of the month (by number) and time of day to run a monthly agent in this dialog box. In addition, you can choose to have the agent run only during a certain period of days. The dialog box also enables you to specify the server on which the agent should run. By default, this field contains the current server. If the agent will need to be run on a different server, this field will need to be updated. If you select the Choose when agent is enabled checkbox, you can select the server when the agent is enabled.

You might choose to use the On Schedule Never option if the agent will only be triggered by an action or other button and not by schedule, event, or from the menus.

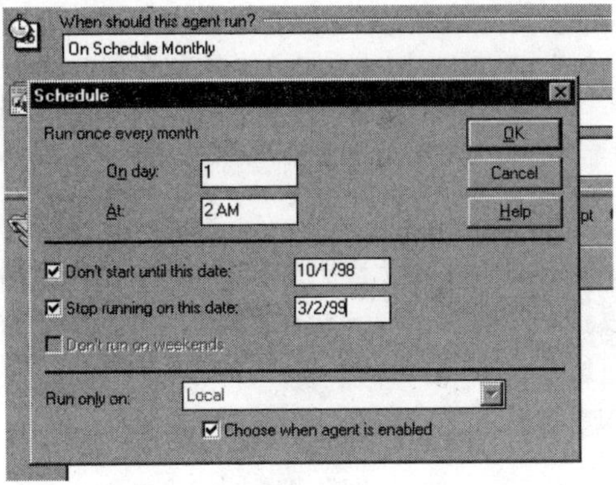

Figure 8.25 Monthly agent options.

The next option in the Agent design window is the choice of which documents the agent should act on. The trigger that you select affects the documents selected. Table 8.8 shows the base selected documents for each trigger. The search builder can then be used to limit the documents selected by the agent.

Table 8.8 The agent triggers enable these documents to be selected

Trigger	Selected Documents
Manual triggers	■ All documents in database ■ All new and modified documents since last run
	■ Run once
If new mail has arrived	■ Newly received mail documents
If documents have been created or modified	■ Newly modified documents
If documents have been pasted	■ Pasted documents
On schedule hourly, daily, monthly, weekly, or never	■ All documents in database ■ All new and modified documents since last run

Again, in addition to the default selection, you can limit or describe the documents to be affected by the agent by using the Search Builder, which is shown in Figure 8.26. Click the Add Search . . . button to display the Search Builder dialog box.

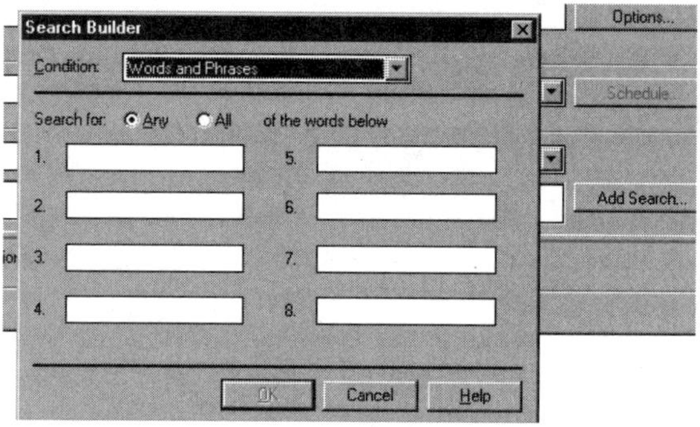

Figure 8.26 Search builder dialog box.

The Search Builder enables you to add conditions to limit the documents. The conditions available from the drop-down list include the following:

- *By Author.* This feature enables you to select documents either created by or not created by a specific user, as shown in Figure 8.27.

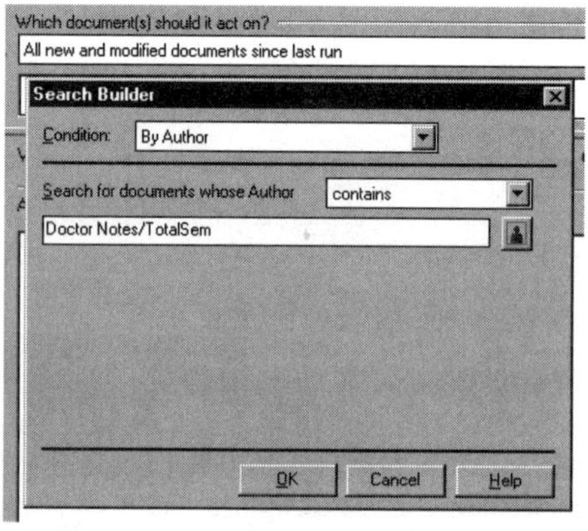

Figure 8.27 By Author search.

- *By Date.* This enables you to select documents based on the date they were either created or modified, as shown in Figure 8.28.

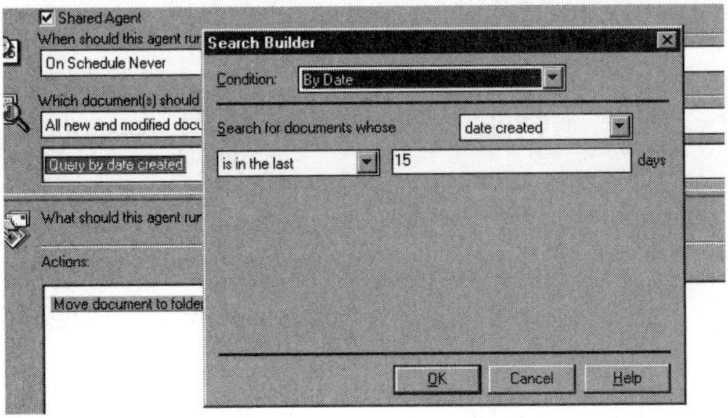

Figure 8.28 By Date search.

- *By Field.* This enables you to select documents based on the contents of a field in the forms. The options on this dialog box, as shown in Figure 8.29, vary based on the field type that you select.

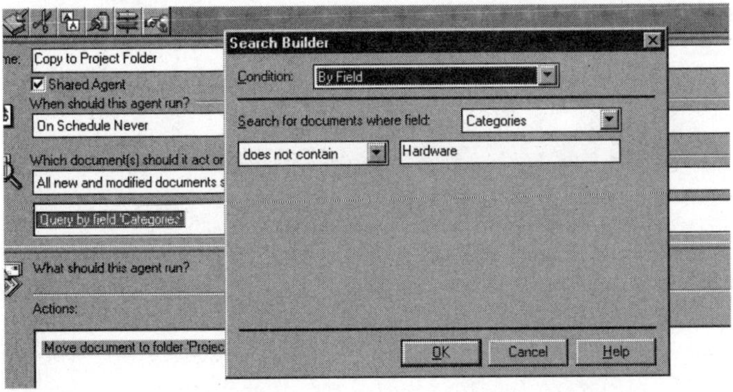

Figure 8.29 By Field search.

- *By Form.* The By Form search condition enables you to select multiple fields on a particular form in which to search. Choose a form, as shown in Figure 8.30, and then add the values to the fields on which you want to base the conditions.

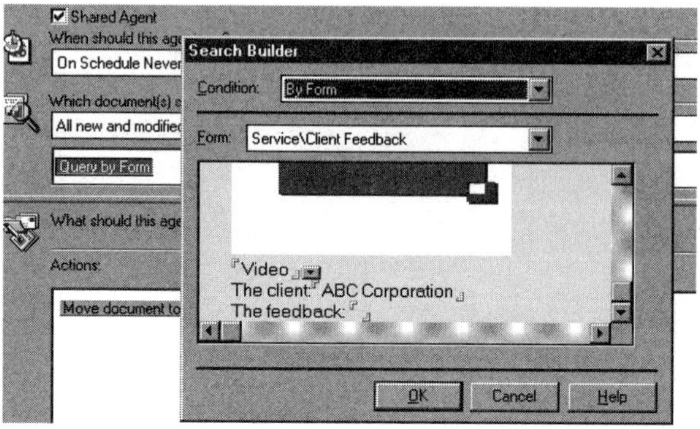

Figure 8.30 By Form search.

- *By Form Used.* Not to be confused with the By Form search, the By Form Used condition enables you to select documents created with a specific form, as shown in Figure 8.31.

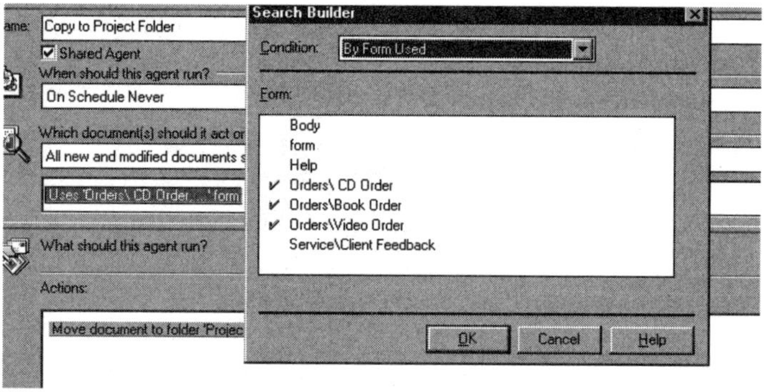

Figure 8.31 By Form Used search.

- *In Folder.* The In Folder search enables you to select documents that are contained in a particular folder, as shown in Figure 8.32.
- *Words and Phrases.* The Words and Phrases search enables you to search for documents that contain particular words or phrases, as shown in Figure 8.33.

After selecting the documents on which the agent should work, the action that the agent will perform needs to be created. Agents can perform actions created by simple actions, @Function formulas, or LotusScript scripts. The following are some of the simple actions available by clicking the Add Action button in this dialog box:

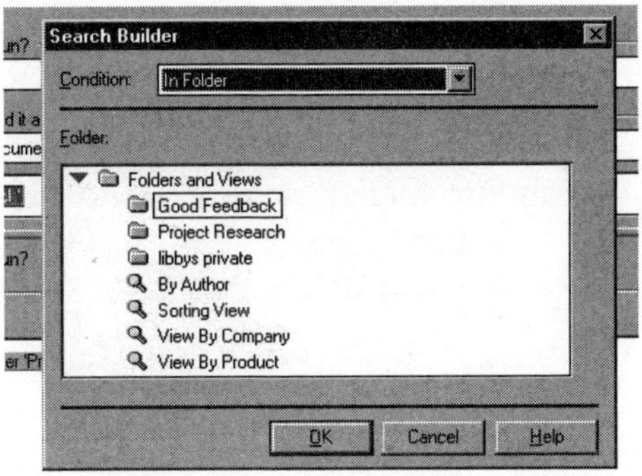

Figure 8.32 In Folder search.

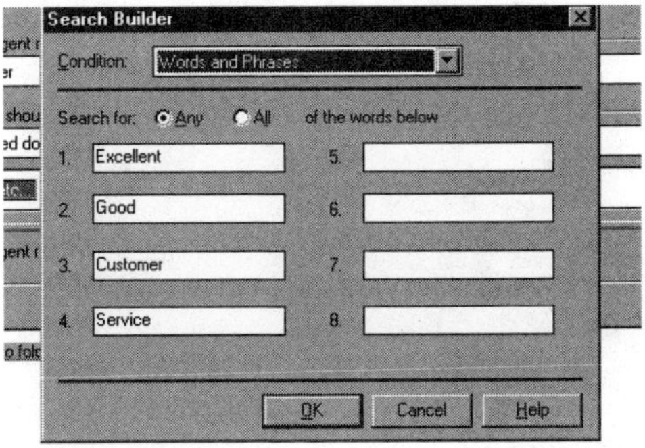

Figure 8.33 Words and Phrases search.

- Copy to database or folder.
- Delete from database.
- Mark documents read or unread.
- Modify field or fields.
- Move to or remove from folder.
- Reply to sender.
- Send document, message, or Newsletter summary.

The Send Newsletter Summary simple action is new to Notes r4. It is likely that this simple action will be tested on the Application Development II exam. The Newsletter summary action will send mail messages that contain doclinks and summaries of the selected documents to anyone you choose. The options for this action are shown in Figure 8.34. You can choose one or more recipients by clicking the person button next to the `To:` field. Then specify a subject for the mail message. To add a `cc:` or `bcc:` field, or to specify these fields or the subject field by formula, click the More . . . button. The body field enables you to type a message to include with the Newsletter summary. You can also include summaries for the selected documents using a specified view. Finally, you can specify a minimum number of documents to be gathered before the messages are sent. See Figure 8.34.

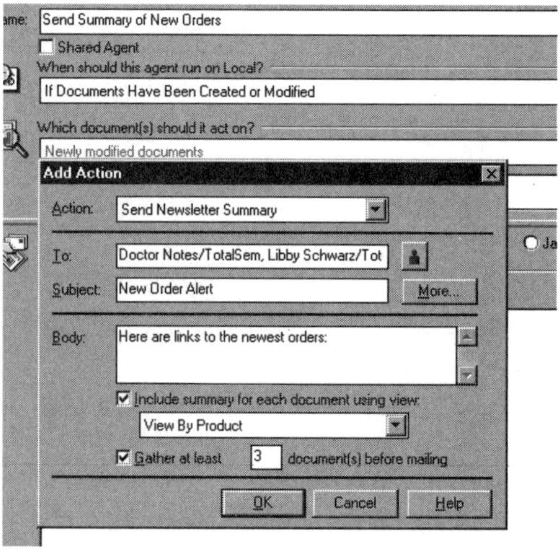

Figure 8.34 Send Newsletter Summary simple action.

You can also use an @Function formula to create an action for an agent to run. If you use an @Function formula, you can choose what the agent does to the documents from the following selections:

- Modify documents.
- Create new documents.
- Select the documents.

By default, Note modifies the documents. There may be times, however, when a new document needs to be created with changes without altering the existing document—or when no changes should be made, such as when testing the selection criteria.

Testing and Troubleshooting Agents

After creating agents using this information, you may need to test or troubleshoot the agents. Two ways to test the agent before running it against live documents include the following:

- Run the Agent in test mode after saving the agent. After you have created and saved the agent, open the Agents view. Select the agent you want to test and choose Actions . . . Test from the menus. The agent will be tested and the Test Run Agent Log dialog box will be displayed, as shown in Figure 8.35. The dialog box will list the number of documents that the agent found to run on and what the agent would have done. This feature enables you to verify that the agent acts as you intended and finds the correct number of documents.

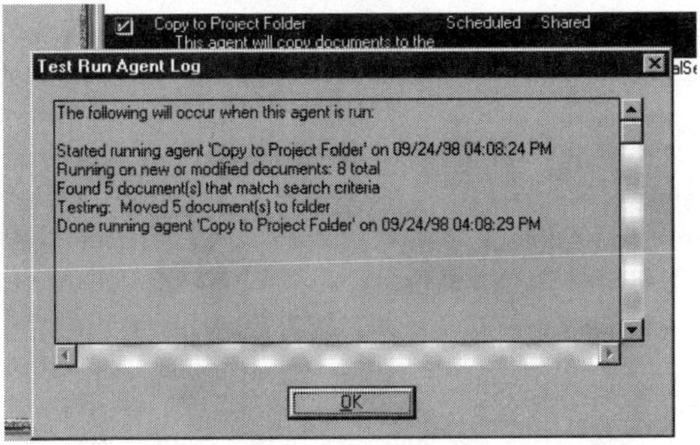

Figure 8.35 Test Run Agent Log dialog box.

- You can also choose to make a copy of the database in which to run the agent. This enables you to run the agent in real mode without affecting production data. After you run the agent in this copy of the database, you can verify that it acted correctly on the documents in the copy.

One way to verify that an agent is running correctly, or to troubleshoot it when it is not, is to use the agent log. The agent log will show data for the last time the agent was run. To view the agent log, select the agent from the Agents view and choose Agent . . . Log . . . from the menus. The Agent Log dialog box, shown in Figure 8.36, will be displayed. Note that it is very similar to the dialog box displayed during the testing of an agent and contains the same type of information.

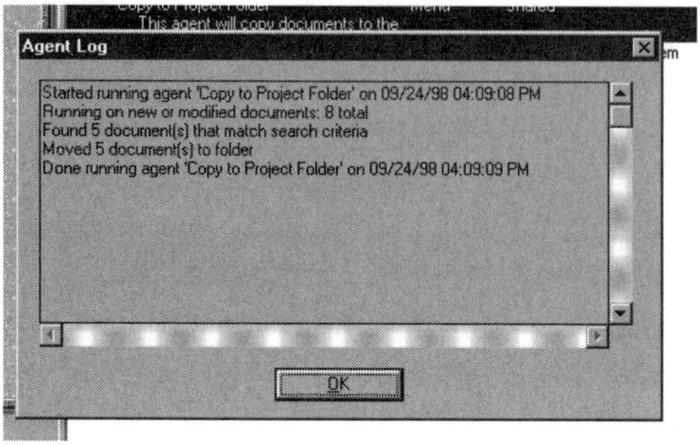

Figure 8.36 Agent Log dialog box.

The following are two issues likely to cause problems with an agent:

- Access
- Agent Manager settings in the server document. The Agent Manager is the server task that runs agents on the server. It is controlled by the server document settings.

An agent runs either under the Access level of the user that created or modified the agent (for scheduled or event triggered agents), or under the Access level of the person running the agent (for agents that are triggered manually). If the person who runs or creates the agent, for example, only has Author access to the database, only documents they created would be affected by the agent, regardless of the selection formula for the agent. Similarly, in the Agent Manager settings of the server document, the Notes Administrator has the ability to define which users can run agents on the server. If the user that creates or runs the agent does not have the right to run agents on the server, the agent cannot run on that server. The Notes Administrator can also limit the times that agents can be run on the server and the number of agents that can run on the server at a time. Verify these settings with the Notes Administrator if the agent is not running as expected.

Creating and Using Actions

Another Notes element that can be used to automate your applications is Actions. Actions can be created in forms or views and can be activated by action buttons on the action bar, by hotspots, or from the Actions menu. Actions cannot run automatically, unlike agents; instead, they must always be triggered by a user.

The process used to create actions was described in Chapter 3, *Fields, Formulas, Agents, and Actions*. To view the actions in design mode, choose View . . . Action Pane from inside any form or view. The Action Pane, as shown in Figure 8.37, will be displayed.

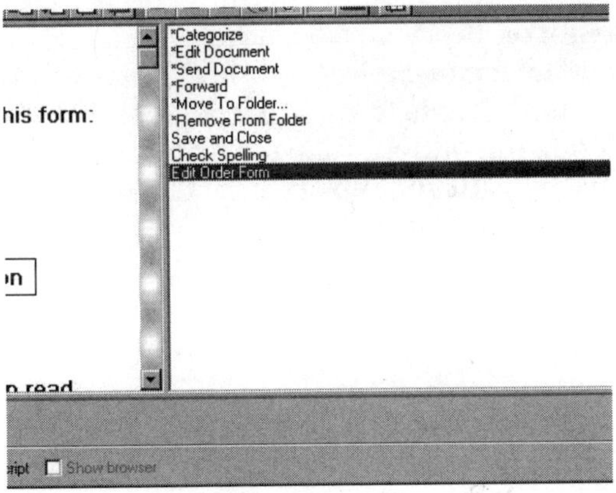

Figure 8.37 Action pane.

When you choose Create . . . Action from the menu when you are in a form or a view, the Action Properties InfoBox is displayed, as shown in Figure 8.38.

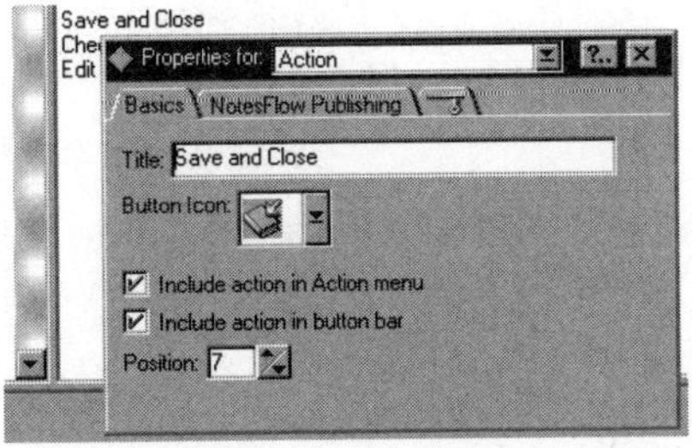

Figure 8.38 Action Properties InfoBox.

You can also choose Create . . . Hotspot . . . Action Hotspot or Button to create an action on a form. Like an agent, the action can run simple functions, @Function formulas, or LotusScript. One use for an action that you are expected to understand is to use an action to trigger an agent. This kind of action can be added by creating an Action (either an Action button or an action hotspot) and choosing to run a simple action from it. When you click the Add Action button, choose Run Agent as the action in the dialog box. The agents in the database will be available from the drop-down list. Select the agent that you want to be triggered by this action. The agent can then be triggered by users when they click the appropriate action button or choose the action from the menu. This feature expands the flexibility of the agents you create.

Actions can also automate menu choices by using @Commands. When you create the action, choose to use a formula instead of a simple action. The @Command button will be available. An example of @Commands that you might run in an action button is shown in Figure 8.39.

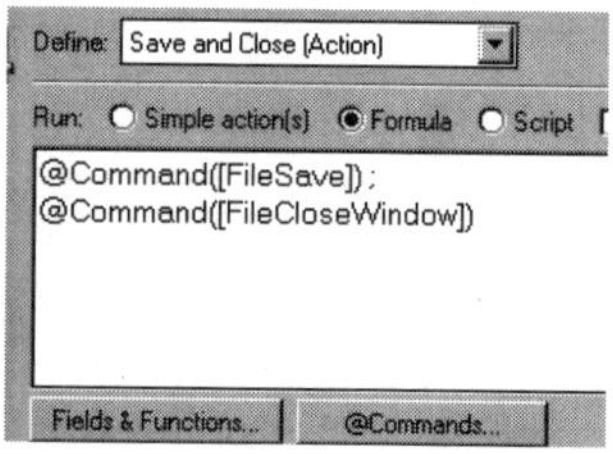

Figure 8.39 @Command action.

NOTE
Only try to use @Commands in formulas if the @Command button is available when you create the formula. If the @Command button is not available, no @Command function is available in this location.

You can also concatenate multiple @Commands and @Functions together in an Action button, to create an advanced formula.

Sample Database

The Support and Training departments have been getting complaints from the managers of other departments. The managers do not know when the PCs are having problems, so they have asked to have e-mails sent to them when a problem or question is sent from their department. Sometimes a manager is away, so the question cannot be automatically sent to a department head. The users need to be able to see all of the department heads and choose to which person they will be sent. To help achieve this, we will create a new field in the Question form called Boss. This field will contain a

@DBColumn formula that the users can use to pick where their copy of the question will be sent.

For the moment, let's get the @DBColumn field created. We will concern ourselves with the mail part later. We can easily add a new form called Depts that has two fields called Department and BName. Next, we make a simple view called Depts with Department as the first Column and BName as the second column. We can then create a @DBColumn formula to reference the Depts view to get a list of managers' names. Use Figure 8.40 as a guide.

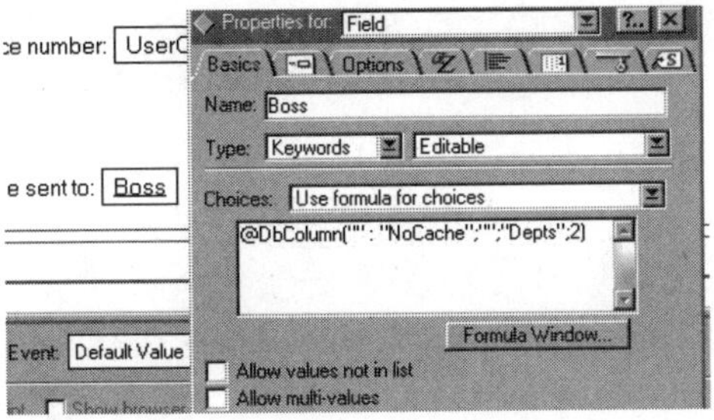

Figure 8.40 Using the @DBColumn function.

Another item to help automate filling out the Question form would be having the Name field filled out with the local username—but giving the user the option of changing the field if desired. For this, we will give the Name field the default value of @UserName. Unfortunately, using just @UserName will give the complete canonical name —far more than we need. To reduce this to the user's simple name, fill out the Default Value for the Name field as shown in Figure 8.41.

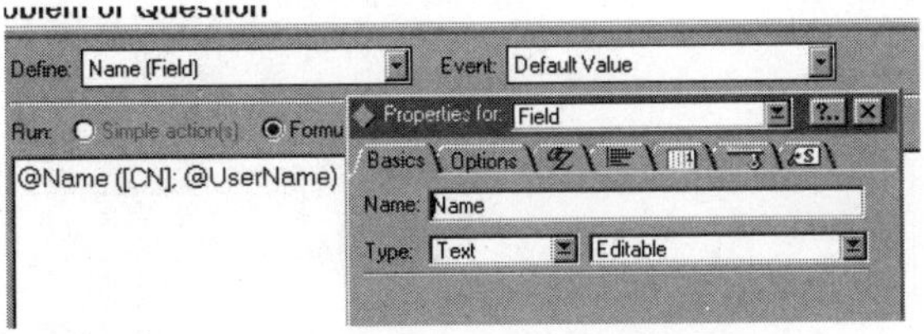

Figure 8.41 Using @UserName and @Name.

Support personnel are having a problem remembering to follow up with users on open calls. We will use the @Prompt function to create a pop-up window that reminds the support personnel to follow up whenever they close a document with the field of Status set to Open. The Close Action that we created earlier will be edited to include the @Prompt. Figure 8.42 shows how the Action Pane has been opened to make the work clearer.

The last item to add to the sample database is an automated Agent that will scan the database periodically and send mail to the two main managers if any problem is more than five days old. Automated agents are actually easy to do—it is simply a matter of telling the agent when to do it. Figure 8.43 shows the settings for the automated Agent. Do not let the long formula intimidate you. It is mostly just text for the @MailSend function.

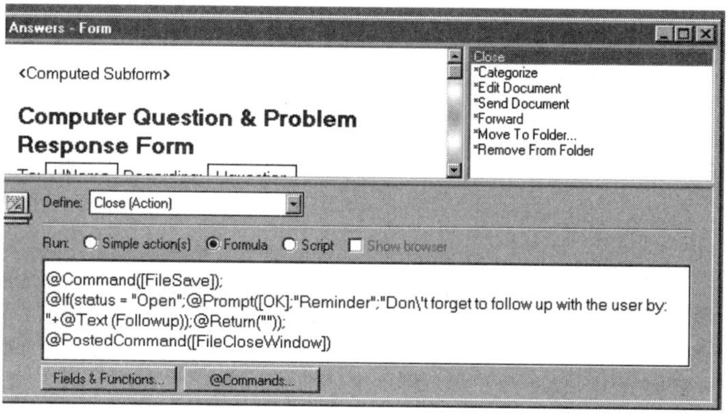

Figure 8.42 Using the @Prompt.

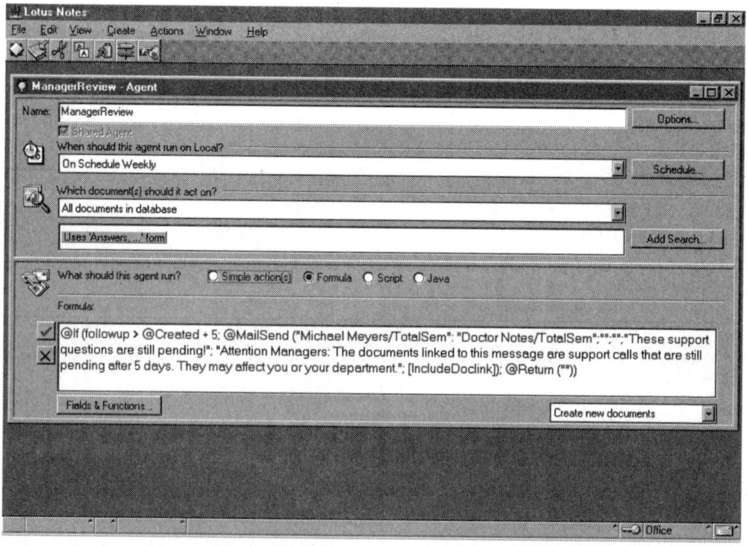

Figure 8.43 Creating the automated agent.

Review Questions

1. Which Notes internal field would tell you who has edited or updated a particular document?

   a. $Editors

   b. $Authors

   c. $UpdatedBy

   d. $Revisions

2. What information does the Notes internal field $Ref give you?

   a. The UNID (Unique Notes Identification number) of the current document.

   b. The UNID of the child to the current document.

   c. The UNID of the parent to the current document.

   d. The UNID of all the descendants to the current document.

3. You can see the values of the Notes internal fields by looking where?

   a. Form Properties InfoBox, Fields tab

   b. Document Properties InfoBox, Fields tab

   c. Add a field to the form with the appropriate name

   d. Database Properties InfoBox, Fields tab

4. David is concerned that users may not be saving the changes that they are making to the documents in the Employee Information database. What can he do to make sure that they save the documents when they make changes?

   a. Create a Note Reserved field called SaveOptions. Set the value to one. All documents will be saved automatically.

   b. Create a Note Reserved field called SaveOptions. Set the value to zero. All documents will be saved automatically.

   c. Create a Note Reserved field called RequiredSave. Set the value to one. All documents will be saved automatically.

   d. Create a Note Reserved field called RequiredSave. Set the value to zero. All documents will be saved automatically.

5. The field Form is an example of what kind of Notes field?
   a. Internal
   b. Reserved

6. Len creates an editable text type field called User with the function @UserName to capture and display the author of the document. It displays the value like this: `CN=Joe Smith/OU=Accounting/O=ABCCorp`. Len also creates a Department field that he wants to populate with the Organizational Unit of the user. How can he do this? Write a formula.

7. Fred creates the following formula in the Input Validation event: `@If (Categories = ""; @Failure ("You must enter a category!"); @Success)`. If they do not complete this field, the users see this message when which events occur?
   a. When the document is saved.
   b. When the document is created.
   c. When the user tabs to the next field.
   d. When the user opens the document.

8. Mandy wants to generate a random text value on a form for a UserID. What @Function can she use?
   a. @Random
   b. @Unique
   c. @UserID
   d. @RandomText

9. A.J. created a form that captures a user's birthday in the Birthday field. She knows that she will need to use this again in other forms and databases. What method can she use to store this for later use, and where will the information be kept?
   a. @SetEnvironment ("UserBirthday"; @Text(Birthday)); stored in the `WIN.INI` file
   b. @Environment ("UserBirthday"; @Text(Birthday)); stored in the `WIN.INI` file
   c. ENVIRONMENT UserBirthday:= @Text (Birthday); stored in the `NOTES.INI` file
   d. @Environment ("UserBirthday"; @Text(Birthday)); stored in the `NOTES.INI` file

10. After setting the environment variable in the question before, A.J. looks in the correct .INI file. What does the environment variable look like?

    a. ENV_UserBirthday

    b. $ENV_UserBirthday

    c. $UserBirthday

    d. $Birthday

11. @DbLookup can be used with Notes databases, Access databases, and FoxPro databases.

    a. True

    b. False

12. Harold wants to allow users to choose a class from a list in a keyword field. This list is fairly long, but already exists in the Class column of the By Class view of the Class Descriptions database. The Class Descriptions database changes daily. What function should Harold use in this field?

    a. `@DbColumn ("": ""; "": ""; "By Class"; 1)`

    b. `@DbColumn ("Notes": "NoCache"; "": "Class Descriptions"; "By Class"; 1)`

    c. `@DbColumn ("Notes": ""; "": "Class Descriptions"; "By Class"; 1)`

    d. `@DbLookup ("Notes": "NoCache"; "": "Class Descriptions"; "By Class"; 1)`

13. The ReplicaID can be used in place of the database name in an @DbColumn or @DbLookup function.

    a. True

    b. False

14. The data limit for an @Db function is:

    a. 128KB

    b. 128MB

    c. 32KB

    d. 64KB

15. The -Default- group on the Class Descriptions database has No Access. Will the @Db function in Question 12 work correctly? If not, what can you do to make it work correctly?

    a. Yes, the @Db function will work correctly.

    b. No, it will not work. You must change the -Default- access to Reader.

    c. No, it will not work. You must give the other database Reader access by typing in the Database Title in the Access Control List

    d. No, it will not work. You must give the other database Reader access by typing in the Database ReplicaID in the Access Control List.

16. The ReplicaID can be used in place of the database name in an @PickList function.

    a. True

    b. False

17. What does the following function do: @PickList ([NAME]: [SINGLE])

    a. Displays the Name view of the current database in a dialog box, enabling the users to pick one name.

    b. Displays the Name view and the Single column of the current database in a dialog box.

    c. Displays the Name and Address book in a dialog box, but only in a single column.

    d. Displays the Address dialog box, enabling the users to pick one name.

18. Which performs a faster lookup, @DbColumn or @PickList?

    a. @DbColumn

    b. @PickList

19. Which of these is not a valid style for the @Prompt dialog box?

    a. OkNoCancel

    b. YesNoCancel

    c. OkCancel

    d. YesCancel

20. The formula @Prompt ([Password]) enables the user to type in the password to their UserID file to open it.

    a. True

    b. False

21. An @Prompt can only return a Yes (1), No (0), or Cancel (-1) value.

    a. True

    b. False

22. Which of the following functions allows a designer to stop evaluation of a formula?

    a. @Stop

    b. @Return

    c. @Goto

    d. @NextFunction

23. Which function is often used in Window Title event formulas?

    a. @IsTitle

    b. @IsDocBeingSaved

    c. @IsNewDoc

    d. @IsDocBeingLoaded

24. Quinn created a private agent to approve time sheets in the time sheet database. When she hires an assistant, she wants to let that assistant use the Approve agent as well. What does she have to do?

    a. Open the Agent Properties InfoBox; enable the Shared option.

    b. Give the assistant Author access to the database and enable the Shared option.

    c. She must recreate the agent as a shared agent.

    d. You can only create private agents.

25. Which of the following are examples of triggers for agents:

    a. A document arriving by mail.

    b. A document being pasted.

    c. A document being saved.

    d. A document being deleted.

26. Why might you choose to have an agent run On Schedule Never?

27. You have created an agent and you want to test its functions without affecting any documents. How can you do this?

    a. Make a copy of the database. Run the agent in the copy.

    b. Use Test Mode to test the agent without running it against documents.

    c. Choose the Test Only option on the Agent Properties InfoBox.

    d. You cannot test the agent without affecting any documents.

Answers

1. c. $UpdatedBy
2. c. The UNID of the parent to the current document.
3. b. Document Properties InfoBox, Fields tab
4. a. Create a Note Reserved field called SaveOptions. Set the value to 1. All documents will be saved automatically.
5. a. Internal
6. @Name ([OU1]; User)
7. a. When the document is saved
8. b. @Unique
9. d. @Environment ("UserBirthday"; @Text(Birthday)); stored in the NOTES.INI file (Note that all the formulas were correct. All the formulas listed would work the same to store an environment variable in the NOTES.INI file)
10. c. $UserBirthday
11. a. True. @DbLookup can be used with both Notes and non-Notes databases.
12. b. @DbColumn ("Notes": "NoCache"; ":"Class Descriptions"; "By Class"; 1)
13. a. True
14. d. 64KB
15. d. No, it will not work. You must give the other database Reader access by typing in the Database ReplicaID in the Access Control List.
16. b. False. You must use the database name.
17. d. Displays the Address dialog box, allowing the users to pick one name.
18. b. @PickList
19. a and d are not valid. The available styles are OK, YesNo, YesNoCancel, OkCancelEdit, OkCancelList, OkCancelCombo, OkCancelEditCombo, OkCancelListMult, and Password.
20. b. False. The Password style of @Prompt does not relate to the user's ID file.

21. b. False. The @Prompt can return a text string or text list also.
22. b. @Return
23. c. @IsNewDoc
24. c. She must recreate the agent as a shared agent.
25. All except d.
26. You would run an agent On Schedule Never if you wanted to have an action or another agent run the agent.
27. b. Use Test Mode to test the agent without running it against documents.

CHAPTER 9

Advanced Views

In Chapter 4, we discussed creating views for the database that display a list of documents based on a selection formula. In this chapter, we will discuss more advanced selection formulas. In addition, when users open documents from the view, Notes has to choose a form with which to display the document. Often, Notes uses the same form that was used to create the document. Sometimes, however, a designer adds a form formula to the view to allow documents to be displayed with a different form. This chapter will describe form formulas and their uses. Other advanced options for views that will be discussed in this chapter include sorting and categorizing columns, especially using hidden columns and allowing users to have input on the sorting. Finally, we will describe advanced column formulas, which enable users to see more information in the views than we discussed previously.

Chapter Objectives

You should be able to answer questions based on the following objectives after reading this chapter:

- Determine the documents that views should display using the search builder and advanced selection formulas.
- Display response and response-to-response documents in a view.
- Display documents in a view using a different form than the one that was used to create the document.
- Allow user interaction with sorting.
- Use hidden columns for sorting in a view.
- Create advanced column formulas using @Functions.

Selecting Documents for a View

In Chapter 4, we described two methods for selecting which documents appear in a view: Search Builder and selection formulas.

NOTE
Note that LotusScript is not an option when creating selection formulas.

To use the Search Builder, click the Add Condition button on the Create View dialog box, as shown in Figure 9.1.

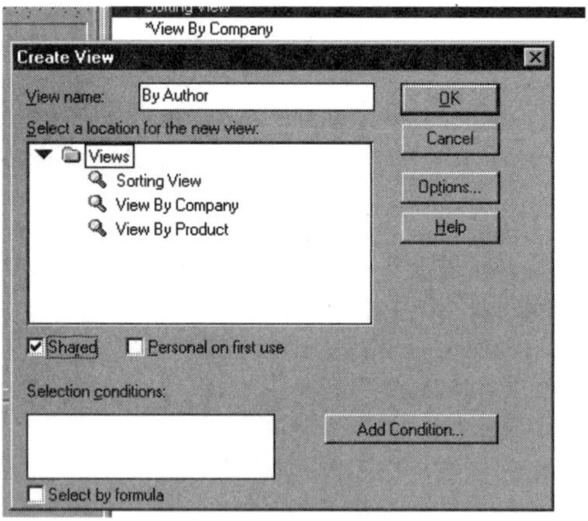

Figure 9.1 Create View dialog box.

The Search Builder allows the designer to select documents for a view based on a variety of criteria. The criteria should seem familiar, as we discussed similar options in Chapter 8 when selecting documents for agents. All of these criteria have more detailed options, as described here:

- *By Author.* This allows you to select documents either created by or not created by a specific user, as shown in Figure 9.2.

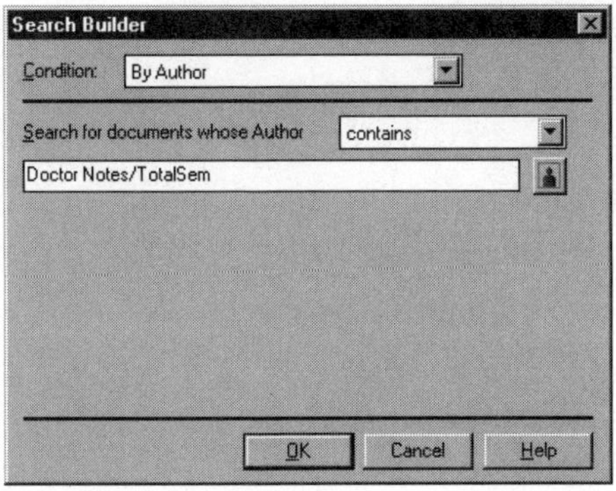

Figure 9.2 By Author search.

- *By Date.* This allows you to select documents based on the date they were either created or modified, as shown in Figure 9.3.

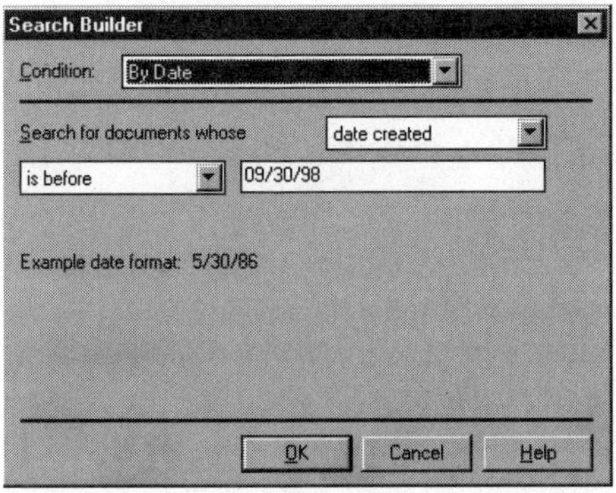

Figure 9.3 By Date search.

- *By Field.* This allows you to select documents based on the contents of a field in the forms. The options here, as shown in Figure 9.4, vary based on the field type that you select.

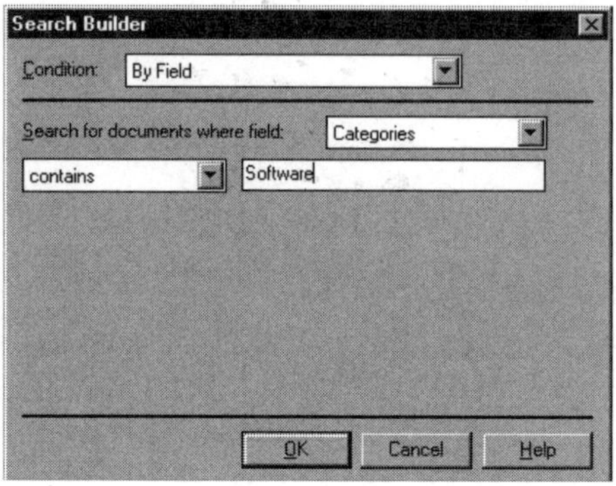

Figure 9.4 By Field search.

- *By Form.* The By Form search condition allows you to select multiple fields on a particular form to search in. Choose a form, as shown in Figure 9.5, and then add the values to the fields that you want to base the conditions on.

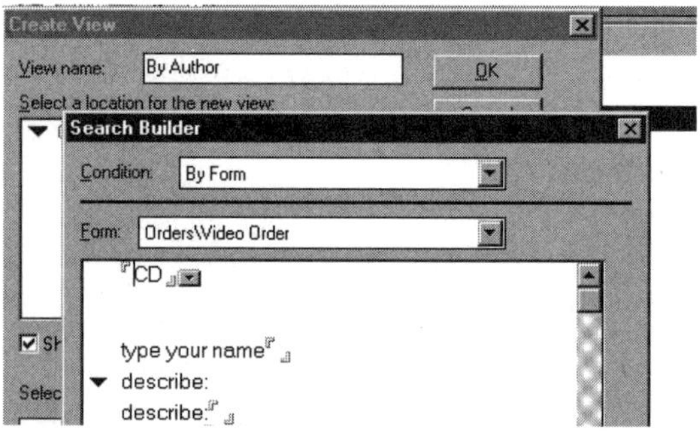

Figure 9.5 By Form search.

- *By Form Used.* Not to be confused with the By Form search, the By Form Used condition allows you to select documents created with a specific form, as shown in Figure 9.6.

Using the Search Builder creates a selection formula automatically, based on the choices you made in the Search Builder dialog box. The choices presented to you in the Search Builder dialog box are the ones that the developers of Notes assumed would be the most used options.

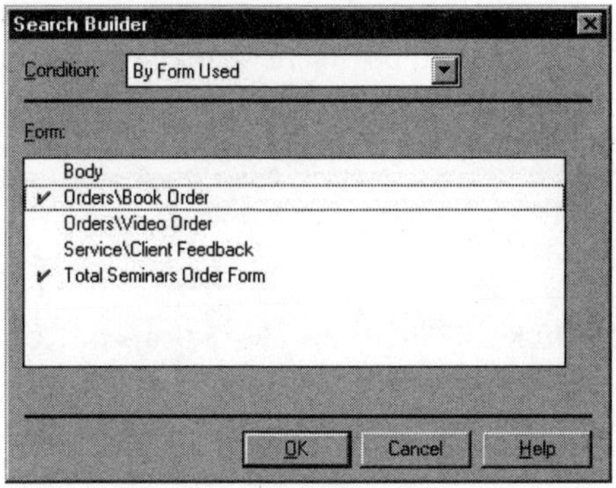

Figure 9.6 By Form Used search.

At times, however, you will need to create a more advanced or complex formula for selecting documents to appear in your views. To create this more complex formula, create a Selection Formula manually. First, click the Select by Formula checkbox on the Create View dialog box. When you select this option, the default selection formula, SELECT @All, is entered into the Selection Conditions field automatically, as shown in Figure 9.7. The default selection formula displays all the documents in the database in the view.

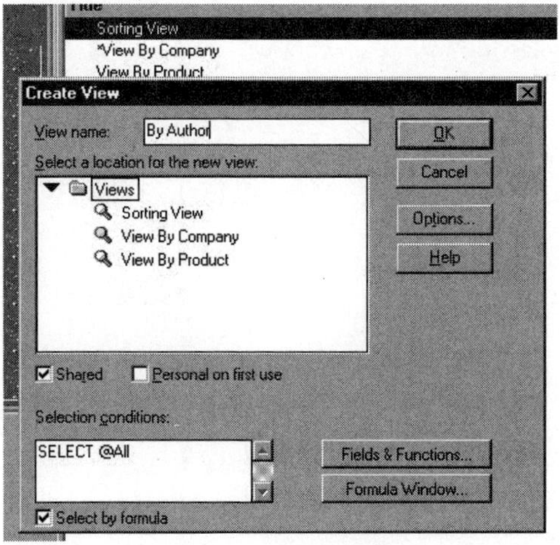

Figure 9.7 Select By Formula.

NOTE
It is often useful to have one view in the database that uses the default SELECT@ALL as the selection formula, just to ensure that no documents get lost. Sometimes developers create selection formulas that are so detailed that some documents do not meet any of the criteria. Having a view, even if it is hidden, showing all documents can be useful in finding lost documents.

You now have the ability to click the Fields and Functions button to add the proper elements to the selection formula. Some examples of selection formulas that might require more than the Search Builder are those that combine forms and field values. If you want to display documents that were created with the Total Seminars Order form, but only by the users Libby Schwarz and Michael Meyers, for example, you would need a formula like this one:

```
SELECT form = "Total Seminars Order form" & OrderTaker =
"Michael Meyers" | OrderTaker = "Libby Schwarz"
```

This formula selects all documents created with the Total Seminars Order form that also have the field OrderTaker completed by Michael Meyers or Libby Schwarz.

NOTE
If you feel you need a larger space to create your selection formula, click the Formula Window button to display a Edit Formula dialog box for the selection formula, as shown in Figure 9.8.

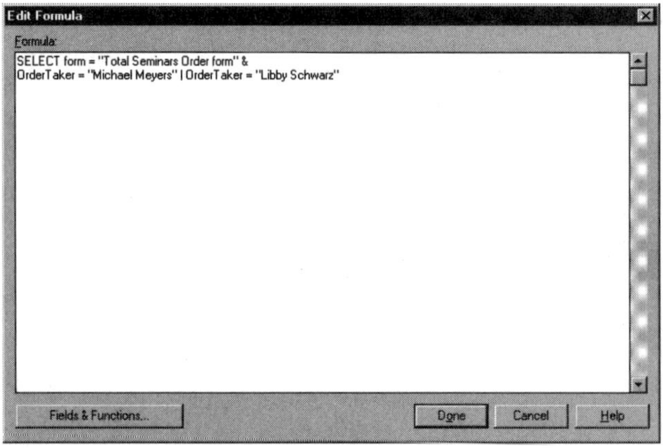

Figure 9.8 Edit Formula dialog box.

Note that the selection formula can provide the same functionality as the Search Builder window. You can create a simple selection formula that selects documents based on the form used, such as `SELECT form = "Total Seminars Order form"` or by the date created, such as `@Created > 09/01/98`.

When creating advanced view selection formulas for choosing documents, some specific @Functions can be used. Two of these essential functions are the @AllChildren and @AllDescendants functions.

@AllChildren is used in both selection formulas and in replication formulas. It selects the documents that are children of the selected document. @AllDescendants is also used in both selection formulas and in replication formulas. This function selects both the children of the selected document and any other descendants.

Children of a document are direct responses to that document. *Descendants* of a document are the responses (children) but also any responses to those responses (grandchildren). There can be up to 32 levels of responses displayed in the view. @AllChildren only displays documents directly related to the selected document, while @AllDescendants displays all the levels of the response hierarchy.

These @Functions are also related to the function @IsResponseDoc. This function displays any document created on a response or response-to-response type of form.

To use these functions, append them to the selection formula to add the response hierarchy to the view, as in the following example:

```
SELECT form = "Total Seminars Order form" | @AllDescendants
```

Both the Search Builder and the selection formula determine which documents will be displayed in the view that you create. The Search Builder allows the selection formula to be built without the designer needing to know or use the formula language. Creating a selection formula from scratch requires more knowledge of the formula language, but offers more complexity and flexibility.

Form Formulas

After designing the view, you may want to choose which form is used to display the documents listed in the view. This is done using a form formula. Form formulas can be used in many ways:

- Allow users to compose documents using one form, but display the data using a different form.

- Create separate views for editing and reading documents. Use different forms for each view.
- Use the data from one form to complete portions of another form.
- Use different forms depending on the user editing the form or the values of certain fields.

By default, Notes determines which form to use to display documents, using the following guidelines:

- When a form is stored in a document, that form is always used to display that document.
- If no form is stored in the document, and a designer has used a form formula for the view, that formula determines the form used to display the documents in the view.
- If no form is stored in the document, and a designer has chosen not to use a form formula, the form used to create the document will be used to display the document as long as it is available. If that form is not available, the default form for the database will be used to display the document.

This section describes creating a form formula. To begin, open the View and open the View Properties InfoBox. Click the Advanced tab (it looks a like a propeller beanie). From the Advanced tab, click the Formula Window button next to the Form formula option, as shown in Figure 9.9.

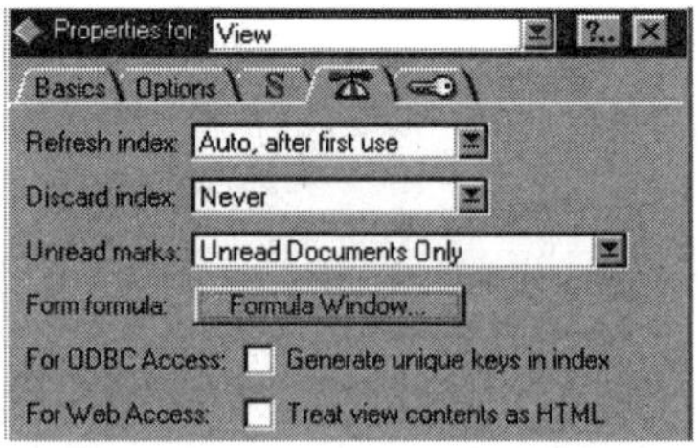

Figure 9.9 View Properties InfoBox, Advanced tab.

The Design Form Formula dialog box is displayed, as shown in Figure 9.10. In this dialog box, you have the ability to create a formula that determines which form is used to display the documents. The formula must result in a Form name. It is common to use an

@If function to determine the appropriate form to use. If you want to use the Completed Order form when an order has been approved, for example, but the Total Seminars Order form any other time, you might use a formula similar to this one:

```
@If(Status = "Approved"; "Completed Order"; "Total Seminars
Order form")
```

After creating the form formula, documents opened within this view will display using the indicated form. Be aware that the form you choose to display the documents must have at least some of the same fields; otherwise, no data will be displayed to the users.

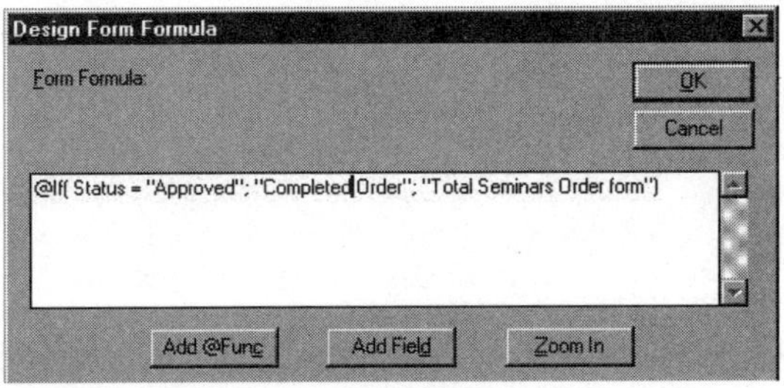

Figure 9.10 Design Form formula dialog box.

Advanced Sorting

In Chapter 4, we discussed creating the columns used to display the data in views and the options available for those columns. One of the options we described was the ability to sort and categorize a view using the Sorting tab on the Column Properties InfoBox, as shown in Figure 9.11.

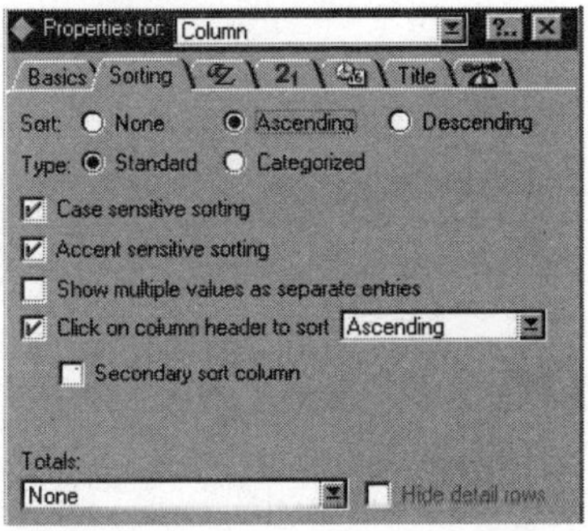

Figure 9.11 Column Properties InfoBox.

The Sorting tab allows the designer to determine the order in which documents are displayed in the view and how those documents are organized. The options available on this tab are discussed in the following paragraphs:

- *Sort.* The Sort option allows the designer to sort documents in ascending order, descending order, or not at all. By default, columns are not sorted. This means that documents are displayed in the order in which they are added to the view. When sorting is enabled, notes sorts in the following order: numbers, letters, accented letters, and punctuation or special characters. Because Notes can sort on one or multiple columns, if multiple columns are sorted, the sorting occurs from left to right.
- *Type.* If the column is sorted, this option determines whether the sort type is Standard or Categorized. The Categorized option, which was described in Chapter 4, allows the designer to group documents together based on the values in a predefined field. Standard sorting is the default type.
- *Case Sensitive Sorting.* By default, sorting is case-insensitive. This option gives the designer the ability to enable sorting based on case as well as on any other column attributes.
- *Accent Sensitive Sorting.* This option allows the designer to sort based on accented letters as well as on any other column attributes.
- *Show multiple values as separate entries.* This option is new to Notes Release 4.6 and will not be tested on the exams.
- *Click on Column Header to Sort.* This option gives the designer the ability to allow the users to sort the data in the view at any time. The user can choose to sort ascending, descending, or both by clicking the up or down arrows that are added to the column headers. The list also contains a Change to view option, which allows the designer to designate a view to be switched to when the user clicks the column header. This option can confuse users, so it is usually a good idea to describe what will happen in the column header (*Click Here to view By Date*). A view that uses these options is shown in Figure 9.12.
- *Secondary Sort Column.* If the Click on column header to sort option has been enabled, this option gives the user an additional sorting column. This refines the sorting order without requiring any user action.

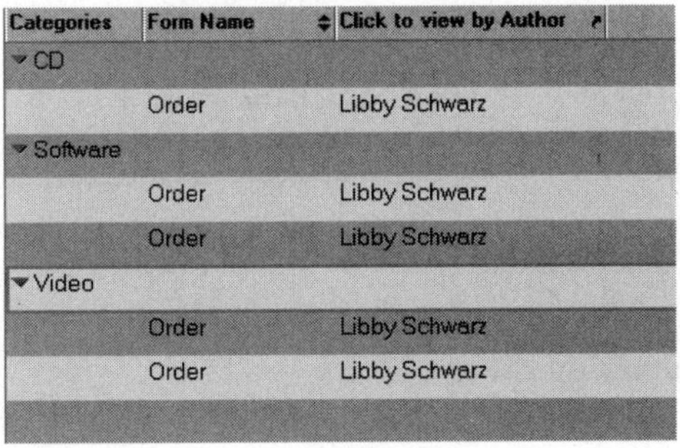

Figure 9.12 View that allows sorting by users.

- *Totals*. This option allows the values in a column to be totaled. It is not tested on the Application Development II exam.

In Figure 9.12, the first column is categorized. The second column allows the user to sort ascending or descending. The third column allows the user to click the column header to go to another view.

The Application Development II exam concentrates on the options that allow the end-user to influence the sorting options. These options are related to the Click on Column Header to sort option.

A further way to manipulate sorting options in views is to create hidden columns. If you want to sort the documents in a view using options that you do not necessarily want displayed in the view, such as month or form, you can create a hidden column and sort on this column.

To create a hidden column for sorting, open the view you are designing in design mode. Create a column and open the Column Properties InfoBox to the Basics tab, as shown in Figure 9.13. After naming the column and adding any other important data to the Basics tab, click in the Hide Column checkbox.

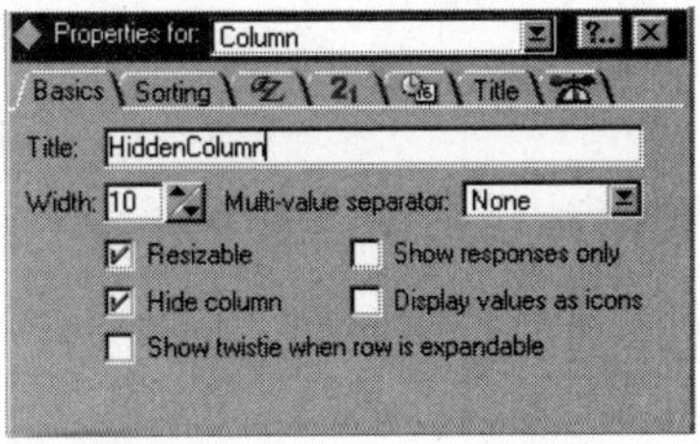

Figure 9.13 Column Properties InfoBox, Basics tab.

NOTE
The column is still visible while you are in design mode; however, when you open the view as an end user would, the column is not visible.

After hiding the column, add a formula for the column using a simple function, a field, or a formula in the design pane. Finally, choose the Sorting tab of the Column properties InfoBox and choose to sort in ascending or descending order.

If you want to sort the documents in the view according to a numeric value of a form, for example, you might write a formula like this one:

```
@If (form = "Book Order"; 1; form = "Video Order" = 2; form = "CD
Order"; 3; form = "CBT Order"; 4; form = "Class Order"; 5; "")
```

This would assign each form a numeric value, one through five. The column would then be sorted according to these values and the documents would appear in order, based on the form that was used to create them.

Similarly, if you want to create a column that sorts based on the month the orders are placed, you could write a formula like this one:

```
@If (OrderMonth = "January"; 1; OrderMonth = "February"; 2;
OrderMonth = "March"; 3; OrderMonth = "April"; 4; OrderMonth =
"May"; 5; OrderMonth = "June"; 6; OrderMonth = "July"; 7;
OrderMonth = "August"; 8; OrderMonth = "September"; 9;
OrderMonth = "October"; 10; OrderMonth = "November"; 11;
OrderMonth = "December"; 12; "")
```

This formula would assign each month name a numeric value that could be sorted in the column.

Advanced Column Formulas

The final step in creating a useful view is determining what is displayed in the columns of the view. As discussed in Chapter 4, a column formula is used to define which data is displayed in the columns. Column formulas can be defined by three things:

- Simple functions
- Fields
- @Function Formulas

Simple functions, as we have seen in other types of formulas, allow the designer to place a value in the field without any complex programming. Examples of simple functions available for display in the view include information about attachments, authors, and the size of the documents, as shown in Figure 9.14.

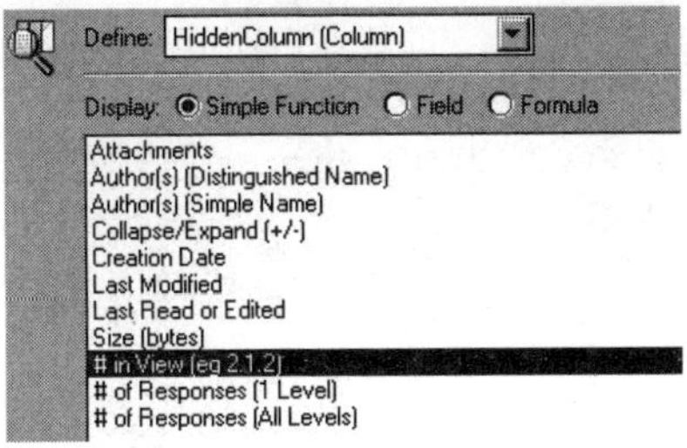

Figure 9.14 Defining a column using a simple function.

When you use a field to determine the contents of a column, you are simply pulling the values from the fields of the documents that are being displayed. Note that if the field you choose is not on all the documents in the view, nothing will be displayed for those documents. You can choose the field to display from the available list. Note that only fields that can be displayed in a view are included in the list, as shown in Figure 9.15. Some field types, such as Rich Text fields, encrypted fields, and Computed for Display fields, cannot be displayed in a view.

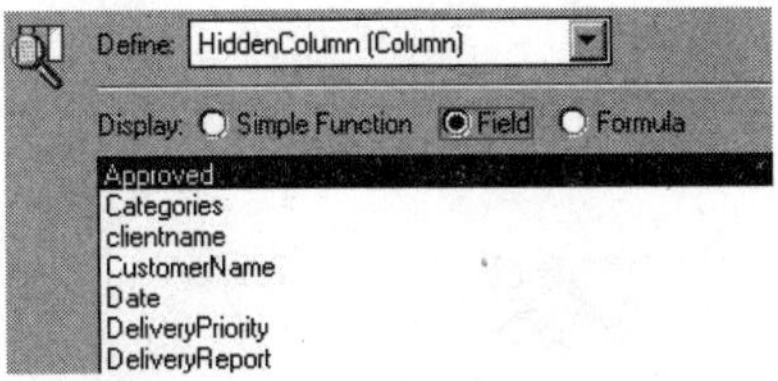

Figure 9.15 Using a field to define a column.

Finally, you can also choose to create an @Function formula to define the contents of a column, as shown in Figure 9.16. This allows you to create a more complex display of data that can include multiple fields, as well as static text or other calculations.

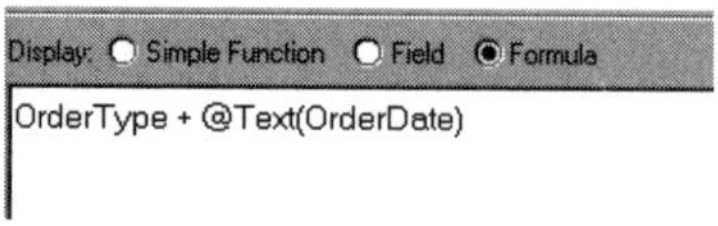

Figure 9.16 Using a formula to define a column.

If you are using more than one field, keep in mind the cardinal rule that only a single data type is allowed in a formula. If you wanted to display the OrderType and the OrderDate in the same column, for example, you might use a formula like this one if the OrderDate field was a Time field type:

```
OrderType + @Text (OrderDate)
```

You may also want to include static text in the column, to explain the data further. If you include static text in a column formula, you must enclose the static text in quotes, for example:

```
"The" + OrderType + " was created on" + @Text(OrderDate) + "by"
+ OrderTaker + "."
```

Any static text is enclosed in quotes, although field names should not be. Keep in mind that the static text will appear exactly the way it is typed inside the quotes, so you should include the appropriate spaces and punctuation.

NOTE

Note that when you change from Simple Action to Field to Formula (or vice versa) on the Design pane radio buttons, you may get an error that warns that *Existing action data will be lost* (See Figure 9.17). This is useful for making sure that you do not accidentally lose the complex formula you have just created.

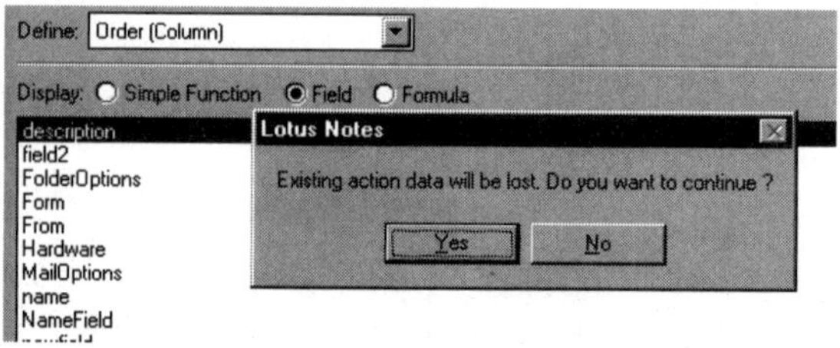

Figure 9.17 Change column definition from simple action to field.

Sample Database

In the last chapter, you added a form called Depts to the database. This form is only used to capture and display information about the departments and their managers and is used to populate the Depts view we used in the @DbColumn formula. Unfortunately, now your other views are suddenly displaying documents that have no information in the columns. This is because the columns are based on fields that do not exist in the Depts form. What we need are some selection formulas! We need one that only shows documents that were created with the Depts form (Depts view) and another one that will prevent documents created with the form Depts from showing up (other views in the database). The selection formulas might look something like the ones shown in Figures 9.18 and 9.19.

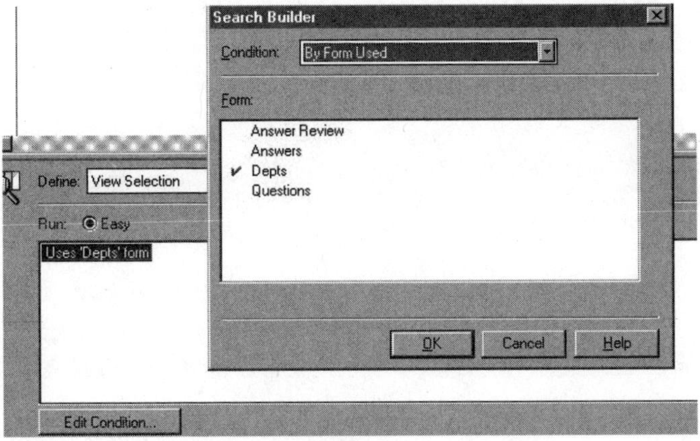

Figure 9.18 Easy Selection formula built with the Search Builder.

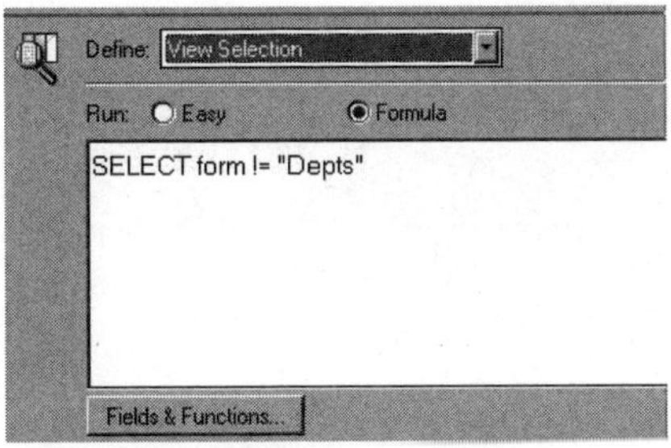

Figure 9.19 Selection formula.

We can use a few advanced formula issues to create a new view that will make the problems easier to track. Let's make a new view called Overdue. Copy the following figures to make the different columns. Notes can really make a view look good!

The Problems column determines and gives the description of the problem (see Figure 9.20).

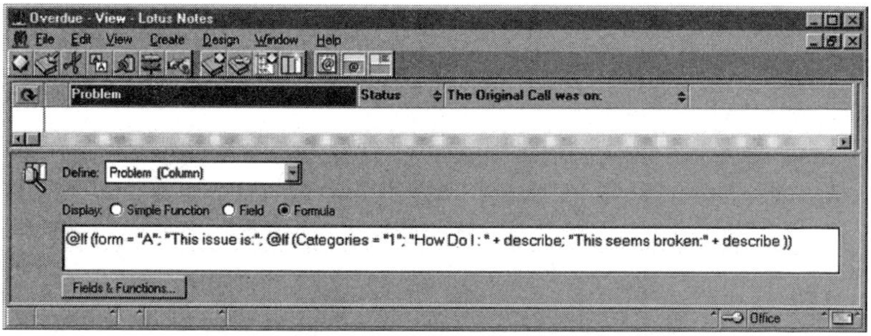

Figure 9.20 Problem column.

The Status column shows the current status of the problem (see Figure 9.21).

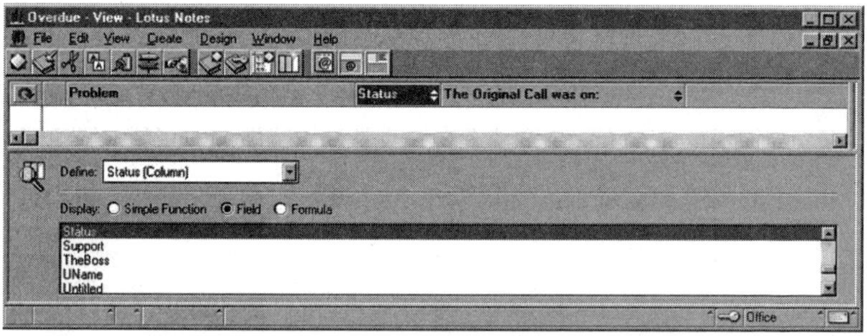

Figure 9.21 Status column.

The Original Call . . . column shows when the problem was first submitted (see Figure 9.22).

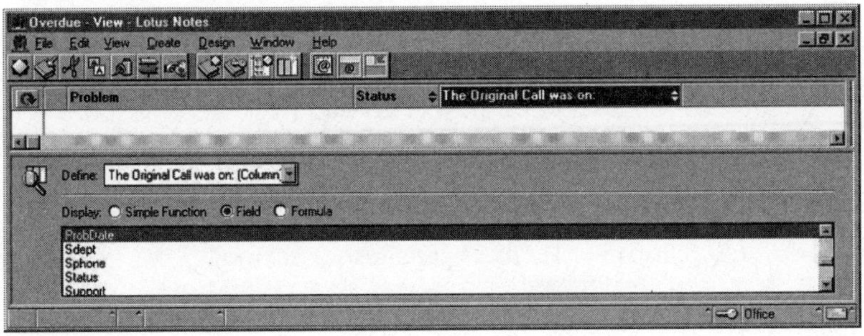

Figure 9.22 Original Call . . . column.

Just for fun, try adding different icons to add stronger visual input. Note how the icons change based on the type of information being shown. Use Lotus Help to get the icon numbers (see Figure 9.23).

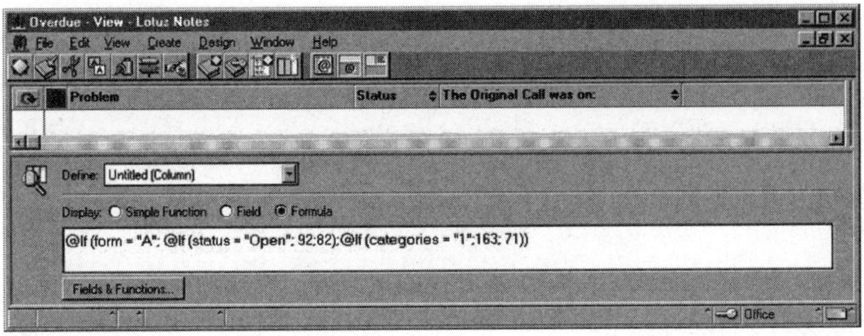

Figure 9.23 Adding changing icons.

Review Questions

1. Which of the following selection formulas would select both documents created with FormA and all of FormA's responses at all levels?

   a. SELECT form = "FormA" & @IsResponseDoc

   b. SELECT form = "FormA" | @IsResponseDoc

   c. SELECT form = "FormA" & @AllChildren

   d. SELECT form = "FormA" | @AllDescendants

2. What is the default selection formula if you do not create one?

3. To select documents based on the values of two fields, would you use:

   a. Search Builder

   b. Selection Formula

4. Peter creates two forms in his database that have many of the same fields. One form captures the data for a company name, address, and contact person. The other form is the basic sales letter for the Video Training product, with the company information ready to be filled in at the top. When Peter opens documents in the Company Profile view, he wants them displayed on the Company Profile form. When he opens documents in the By Letter Sent view, he wants them to be displayed on the Sales Letter form. How can he do this?

5. Jim creates a form called Profile that uses the option to store the form in the document. He creates a By Profile view and a By Company view. The By Company View uses a form formula that selects the Company Info form for displaying documents. The selection formula for both views is SELECT @All. Which form will Notes use to display the documents in when the documents are opened from the By Company View?

   a. Company Info Form

   b. Default form for the database

   c. Profile

6. What does the @Picklist ([NAME]) display in a column in the By Name view?

   a. All the names form the Address book
   b. One column from the Address book
   c. A dialog box to choose names from the address book

7. Alison has two views in the database she is creating: By Company and By Order. When her users are in the By Order view, she wants them to be able to open the By Company view. What option can she use to do this?

   a. She can choose a Secondary View option in the View Properties InfoBox.
   b. She can choose a Secondary View option in the Column Properties InfoBox.
   c. She can choose Click on Column Header to Change to view option in the View Properties InfoBox.
   d. She can choose Click on Column Header to Change to view option in the Column Properties InfoBox.

8. A user calls the help desk and asks what the blue triangles mean in her view. When you ask which blue triangles, she tells you the ones at the top. What do the blue triangles do?

   a. Sort the column
   b. Collapse the column
   c. Expand the column
   d. Close the view

9. How do you create a hidden column? Why would you hide a column?

    a. Place an asterisk at the end of the name; hide the column for security reasons.

    b. Choose the Hide Column option in the Column Properties InfoBox; hide the column for sorting purposes.

    c. Place the word [Hidden] at the end of the name; hide the column for security reasons.

    d. Choose the Hide Column option in the Column Properties InfoBox; hide the column for security reasons.

10. How could you sort on the field ApprovalStatus in status order instead of in alphabetical order? The choices in the ApprovalStatus field are Approved, Rejected, Pending, and Incomplete.

Answers

1. d. SELECT form = "FormA" | @AllDescendants.
2. SELECT @All.
3. You could use either A or B. To use the Search Builder, choose to select By Form.
4. He can use a form formula in the views to choose which form to use to display the documents.
5. c. Profile.
6. None of the above. You cannot use @Picklist in a column or view. It displays a column of a view.
7. d. She can choose Click on Column Header to Change to view option in the Column Properties InfoBox.
8. a. Sort the column
9. b. Choose the Hide Column option in the Column Properties InfoBox; hide the column for sorting purposes.
10. Create a hidden column, sorted in Ascending order, with the following formula @If (ApprovalStatus = "Approved"; 1; ApprovalStatus = "Rejected"; 2; ApprovalStatus = "Pending"; 3; ApprovalStatus = "Incomplete"; 4; "").

CHAPTER 10

Maintain Design Consistency

As you begin creating advanced applications for larger environments and sharing the tasks of designing databases with other developers, you may want to automate the tasks of design management. *Design management* is the process of maintaining and updating the design—both the appearance and function—of a database and it is best accomplished by using design templates. A *design template* is either a Notes template (.NTF) or a Notes database (.NSF) that stores the design elements that a developer has created. It then allows databases to inherit the changes made to these design elements. In addition, design templates, as we discussed in Chapter 1, can assist a designer in creating a new database, as either the entire database or some elements of the database can be based on the elements in the design template. This can be done when creating the database, using methods we discussed in Chapter 1, or it can be done by copying elements from the template, such as views or forms, into a new database.

Design development and maintenance require three basic skill sets, covered in order in this chapter. First, you must create and define a design template. Second, you must know how to update the design elements after a developer has made changes to the template. Finally, you must understand the process required to determine which elements of a database inherit their design from the design templates. Mastering these three skill sets will make you a solid Notes developer (and help a lot on the Application Development II Exam).

Chapter Objectives

You should be able to answer questions based on the following objectives after reading this chapter:

- Create a design template.
- Use a design template to create a new database.
- Refresh or replace the design of a database using a design template.
- Define the elements that should inherit information from a particular design template.
- Force elements of a database not to inherit design information from a template.

Create and Define a Design Template

A design template can greatly facilitate the development of Notes databases in larger environments by automating some design features and creating a degree of standardization. Specifically, templates have three great advantages in such an environment. First, you can make changes to the design of the application without interrupting the use of the application. Second, you or another designer can create new databases that appear consistent in design to the end users. Finally, you can save design time when creating similar databases because you can place all the shared design features on the template.

Two types of templates exist in a Notes' world: a Notes template file (.NTF) and a design or master template. A *Notes template file* is a simple template that contains the basic outline of a database. Notes template files are very useful for creating a series of new databases that share common features. These templates may not link dynamically to other databases and are therefore not capable of updating any database design features after a database has been created. A *design template* or *master template* (the terms are synonymous and interchangeable on the Application Development II Examination) contains the basic outline of a database for development, but *can* link dynamically to other databases. This grants the developer the ability to make changes to the master template that automatically update all the databases built on and linked to the design template.

NOTE

Note that most of the templates that are available by default within Notes are design, or master, templates that can be linked dynamically to databases. Databases, such as user mail files and the Public Name and Address book, are linked to the design templates used to create them and will have their design updated automatically. This can be an issue if a decision is made to customize these databases. Either the customization must be done at the design template level, or the database must stop inheriting its design information from the template.

The template creation process differs somewhat between the two different types of templates. Both require that you first build a complete database. The templates are then based on this database. To create a Notes template file, make a copy of the completed database, using File . . . Database . . . New Copy. Instead of making the file a .NSF file, use the .NTF extension to indicate that the file is a Notes template. Also, choose to copy the database design only. This creates a template file that you can use as a base when creating new databases. It will be listed in the Template list on the New Database dialog box.

To create a design, or master, template that can be dynamically linked to other databases, open the Database Properties InfoBox. Click the Design tab and enable the Database is a Template checkbox, as shown in Figure 10.1. You must also give the template a template name. This template name should have no spaces and should describe its function.

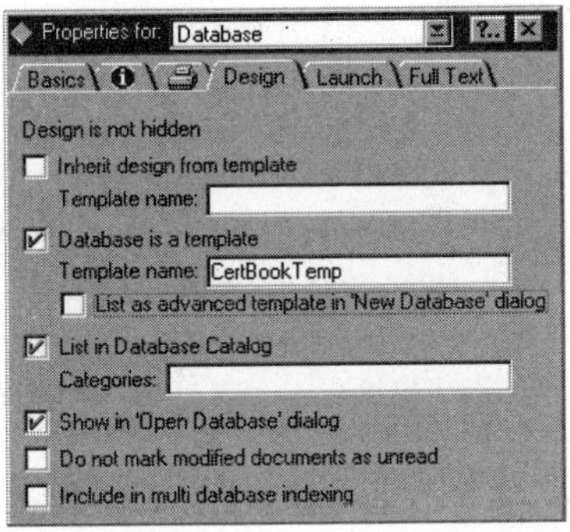

Figure 10.1 Database is a template.

NOTE

The Database is a Template **property is the defining characteristic of a master or design template, as opposed to a Notes template file. When** ***The Database is a Template*** **property is enabled, design can be automatically updated in the databases using the template. If this option is not enabled, only a design replace process can be used to update databases with design changes.**

After creating a master template, you will want to link databases to it so that they can inherit the design information of the template. To link a new database to the master template as you create the database, enable the Inherit future design changes checkbox on the New Database dialog box, as shown in Figure 10.2. Once you create the database using these options, the design template is dynamically linked to the database, and design refreshes can occur automatically.

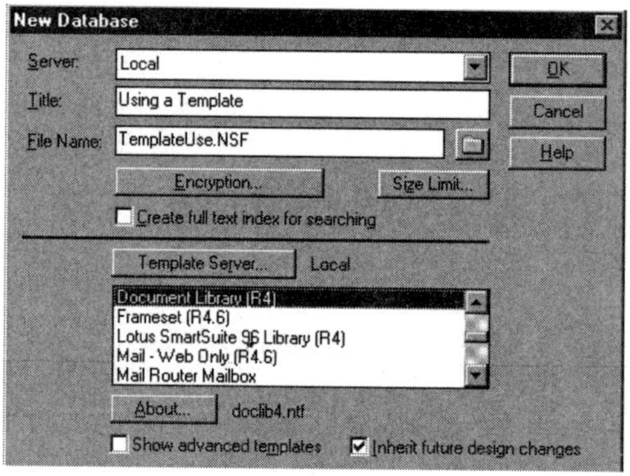

Figure 10.2 Inherit future design changes.

If, however, you want to link a design template to an already existing database, open the Database Properties InfoBox for the database that you want to link to the template. On the Design tab, enable the Inherit Design from Template checkbox, as shown in Figure 10.3. Type in the name of the design template from which you want the database to inherit design elements. Spelling or capitalization differences disable the inheritance from the template. It can be useful, therefore, to open the Database properties of the design template and copy the template name, using Ctrl + C. You can then paste the template name into the database properties of the database using Ctrl + V.

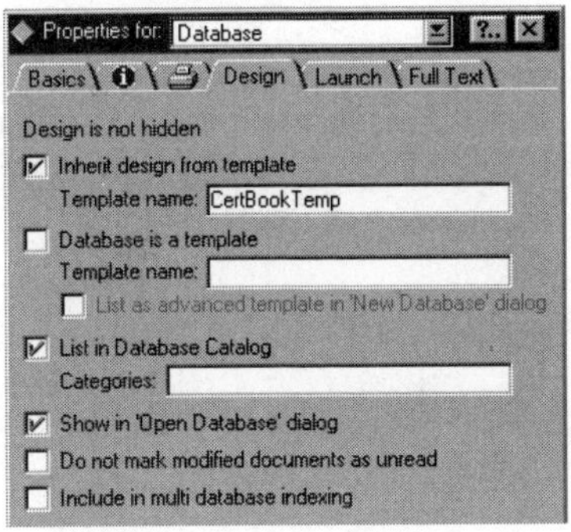

Figure 10.3 Inherit design from template.

NOTE

When you have defined a database to inherit design elements from a template, you will get an error if you try to edit those design elements, as shown in Figure 10.4. This is to remind you that when the design of the database is refreshed from the design template, any design changes made directly to the database will be lost.

Figure 10.4 Design change error message.

Refreshing the Design of a Database

When you make changes to the design elements of a template, you will need to push those changes out to the databases that were created with or are linked to the templates. These changes can be done automatically, using a server task, or manually. The following are the methods of distributing design changes to databases:

- Automatically, using Server Design task
- Manually, using File . . . Database . . . Refresh Design
- Manually, using File . . . Database . . . Replace Design

When you link a database to a design template dynamically, any changes that you make to the design template will be passed on to the database automatically. A server task called Design runs on Domino servers every day at 1 a.m. by default. The Design server task refreshes the design of the databases that are set to inherit design from a design template automatically. To verify that the Design task runs on your server, open the server's NOTES.INI file, as shown in Figure 10.5, and look for the ServerTasksAt1 line. The Design task should be listed on this line.

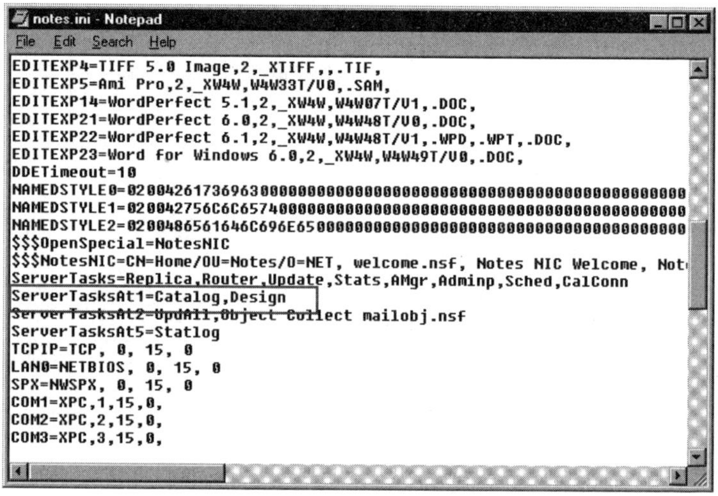

Figure 10.5 Server's NOTES.INI showing the Design task.

NOTE

Be aware that for the Design task to refresh design elements automatically, the design templates must be available in the \NOTES\DATA directory of the server. In addition, the template must be enabled for Replication. To verify that replication is not disabled for the template, right-click the database icon and choose Replication settings. Make sure the checkbox for Temporarily disable replication is *not* selected, as shown in Figure 10.6.

When a design template is stored on another server or in another folder, you will need to update the design manually. One way to do this is by refreshing the design by choosing File . . . Database . . . Refresh Design. You will then choose the appropriate server, as shown in Figure 10.7, based on where the design template is located. When a database's design is refreshed, any design changes made to the chosen design template are passed on to the database. Refreshes are incremental. Be aware that any design elements that are marked

Do not allow replace/refresh to update will not refresh, even if the design element has been updated in the design template.

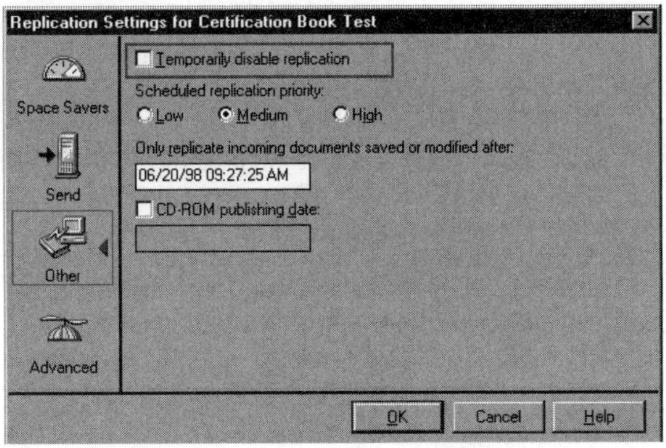

Figure 10.6 Replication settings dialog box.

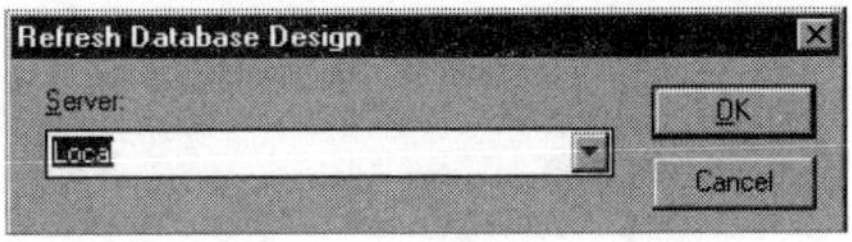

Figure 10.7 Refresh database design dialog box.

When a database is not linked to a design template using the Inherit Design from Template option, you can replace the design elements of the database with those from a design template by using File . . . Database . . . Replace Design. Replacing the design replaces the design elements of the database with those of the design template completely and indiscriminately. If there are any customized views or folders in the database, for example, they will be lost. Replacing the design makes the database identical to the template, with the following exceptions of elements that are not replaced during a Replace Design process:

- Database icon
- Database title and category
- Most of the database security, including the database Access Control List (ACL) and any encryption settings

- The About this Database and the Using this Database documents
- Any elements that are marked Do not allow replace/refresh to update

You can also control the inheritance of certain design elements separately. You have the option either to force the specific design elements to inherit their design from a design template or to force the design element not to be refreshed during design refreshes or replacements. Both of these options are activated using the design document for a design element. The following elements can be controlled using their design documents:

- Forms
- Subforms
- Views
- Shared fields
- Navigators
- Folders
- Agents

To view the Design document for a particular design element, select that design element from the appropriate design view. Choose Edit . . . Properties from the menus. A sample of a design document is shown in Figure 10.8.

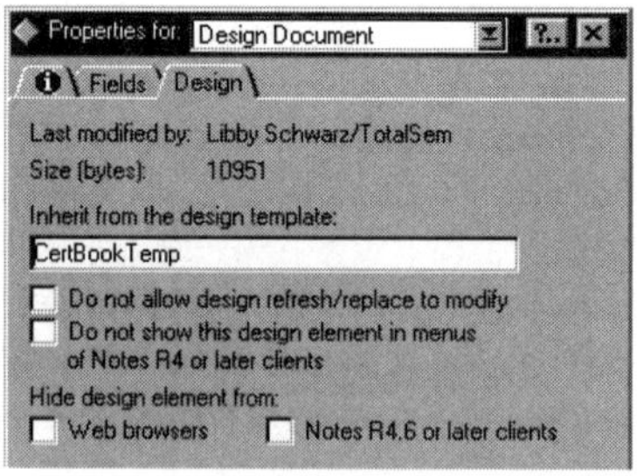

Figure 10.8 Design document.

NOTE

You see a different dialog box when you access the properties from a view rather than from within the design element. If you open a form and then choose Edit . . . Properties, for example,

Notes displays the Form Properties InfoBox. However, if you select the form from the Design . . . Forms view and choose Edit . . . Properties, Notes displays the design document.

After you open the design document, you have the following important options:

- *Inherit from the Design Template.* This option selects a design template for this particular design element. Type in the name of the design template, or copy and paste it from the Database Properties InfoBox of the template. This design element will be dynamically linked to the design template specified.
- *Do not allow design refresh/replace to modify.* Use this option if you want to protect a design element from being changed by design refresh or replace processes. Selecting this freezes this design element unless you remove the option and allow the design to refresh or replace. This option is shown in Figure 10.9.

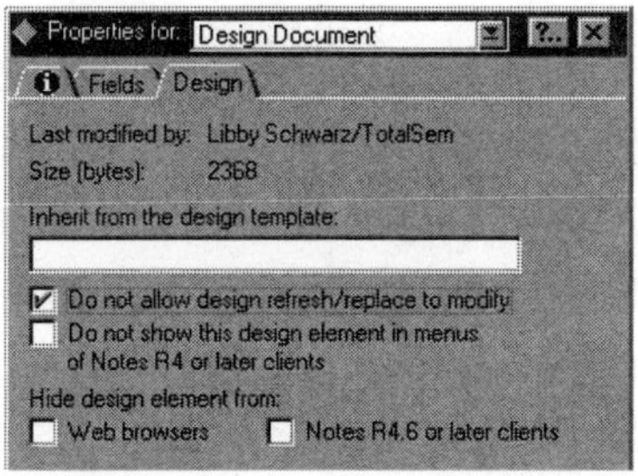

Figure 10.9 Do Not Allow Design Refresh/Replace to Modify.

Sample Database

After you have created this database, you realize that it will soon be time to put it out on the server and allow people to start using it. You also know that as soon as you do so, they are going to want to request changes and additions. You decide to make a copy of the database to use as a design template, so that when you need to make changes, you can do so without disrupting the use of the database.

First, make a copy of the database, as shown in Figure 10.10. Remember that you only want to make a copy of the design elements, not any of the data. You will also want to use a .NTF extension for the template.

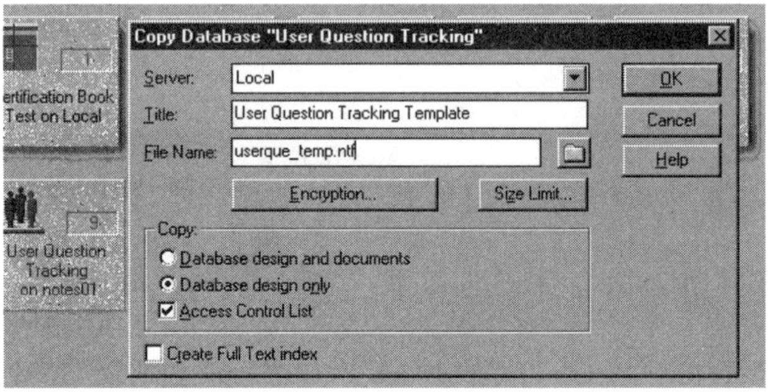

Figure 10.10 Make a copy of the database.

Next, define the template file as a design template, using the Database Properties InfoBox for the template, as shown in Figure 10.11.

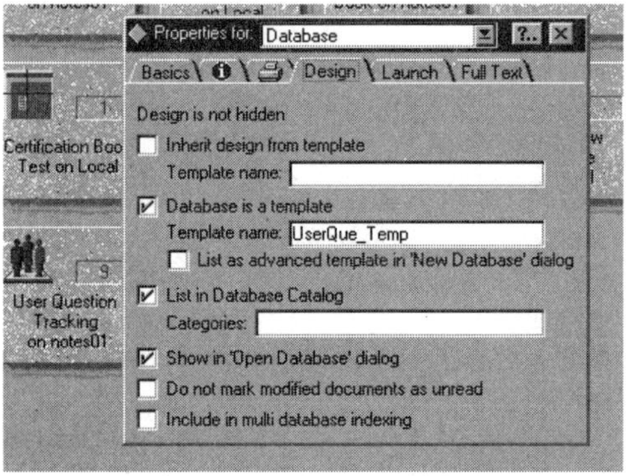

Figure 10.11 Define database as a design template.

Before the database will automatically inherit the changes you make, you have to tell it which template to use to inherit the changes. This is done on the Database Properties InfoBox for the database, as shown in Figure 10.12.

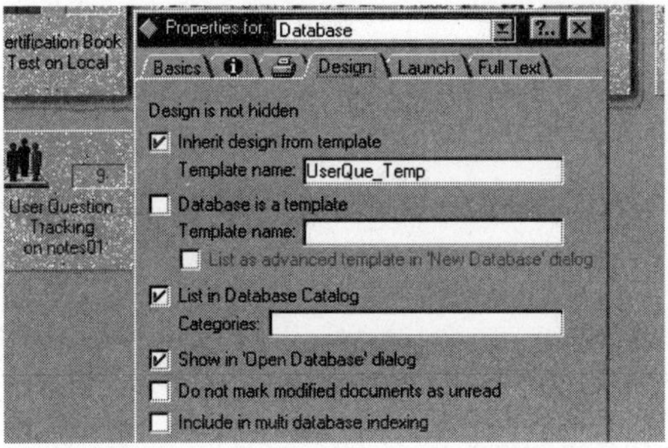

Figure 10.12 What template do I use for inheritance?

All design elements in the database will now inherit design changes from the design template. However, you may want to make some design changes that should not be passed directly through. The Depts form and Depts view, for example, are going to be changed, but you are not sure that you want your changes being populated to the production database. To lock out refresh changes for these design elements, use the design document for the element in the database, as shown in Figure 10.13.

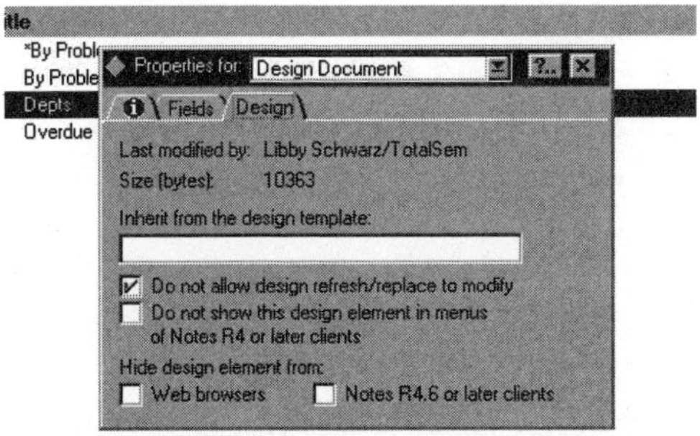

Figure 10.13 Use the design document to prevent changes.

Review Questions

1. The DESIGN task runs on the server at 1 a.m. What does it do to the design of the databases?

    a. Refresh

    b. Replace

2. When making a new design template out of a completed database, what steps must you complete?

    a. Copy the database, including all design elements and documents.

    b. Name the file with a .NTF extension.

    c. Enable the Database is a Template property in the Database Properties InfoBox.

    d. Choose to List as Advanced Template in the New Database dialog box.

3. What is the difference between a template and a master or design template?

    a. There is no difference. Those are all different names for the same thing.

    b. The master or design template has to have a .NTF extension.

    c. The master or design template can have either a .NTF extension or a .NSF extension.

    d. The master or design template has the database as a template property.

4. How do you link a database to a template?

    a. Choose Inherit Future Design changes on the New Database dialog box.

    b. Choose Inherit Future Design changes on the Database Properties InfoBox.

    c. Choose Inherit Design from Template on the New Database dialog box.

    d. Choose Inherit Design from Template on the Database Properties InfoBox.

5. Robert creates a database based on the Discussion template. He chooses Inherit Future design changes when he creates the database. Later he decides that he needs to edit one of the forms to add a field. Can he do this?

   a. Yes.

   b. No.

   c. Yes, but he will get an error first.

   d. Yes, but the changes will be overwritten by the DESIGN task.

6. These elements are refreshed when the DESIGN task runs on the USERPROFILE.NSF:

   a. Database Icon

   b. About this Database Document

   c. Shared Fields

   d. Database Title

7. Harry created a database based on the Discussion template provided with Notes. Now he wants to make changes to one of the forms and wants to make sure that those changes are not overwritten. He still wants the rest of the design to be refreshed nightly. What can he do?

   a. When he makes changes to the form, he will have to break the link to the Design template.

   b. He will have to create an agent that makes the changes every day. The DESIGN task will always overwrite his changes.

   c. He can use the design document for the form to choose Do not allow design refresh/replace to modify.

   d. He can use the Form Properties for the form to choose Do not allow design refresh/replace to modify.

8. Louise wants to copy a few forms from the Room Reservations template provided with Notes and put them in her new Asset Tracking database. She wants those forms to continue to be updated if the Room Reservations template changes. How can she do this for those forms without having the rest of the database based on and updated by the Room Reservations template?

Answers

1. a. Refresh.
2. b and c are correct. Name the file with a .NTF extension and enable the Database is a Template property in the Database Properties InfoBox.
3. c and d are correct. The master or design template can have either a .NTF extension or a .NSF extension and the master or design template has the Database is a template property.
4. a is correct when you are creating a new database based on a template and you will want it to continue to inherit design changes from that template. d is correct when you are attaching a template to an already created database.
5. c and d are the best answers. He will get an error when editing a design element that is tied to a design template and the changes can be overwritten when the design is refreshed.
6. c. Shared Fields is the only correct answer. The other elements are not refreshed or replaced automatically.
7. c. He can use the design document for the form to choose Do not allow design refresh/replace to modify.
8. She copies the forms into her new database. She opens the design document for the forms and chooses the Design tab. She puts the template name for the Room Reservations template in the Inherit from the design template field.

CHAPTER 11

Security in Applications

Lotus Notes provides a wide range of security options to protect your applications and data. These security options begin at the authentication and server access level and continue down through the view, form, document, and field levels. In Chapter 5, we described one of these levels of security, the *Access Control List* (ACL), although only in the most limited detail. This chapter describes the levels of security in more detail, including how to implement them. It concentrates on the levels of security that you would be expected to implement and administer as an application developer. It also describes the other levels in at least limited detail to help you understand how the levels of security work together.

Chapter Objectives

You should be able to answer questions based on the following objectives after reading this chapter:

- Describe the levels of Lotus Notes security.
- Implement appropriate ACL security at the application level.
- Add Roles to an ACL and utilize them in the security of the forms, views, and documents of the database.

- Use Form and View access lists to refine the security in a database.
- Implement Readers and Authors fields in conjunction with the ACL of a database for enhanced security.
- Add signing and encryption to fields and forms.

Lotus Notes Security

Security in a Lotus Notes environment actually begins with security outside the software environment altogether. To secure your server, applications, and data, the security should begin at a hardware level. First, place the servers and workstations in your environment in locked areas to prevent physical access. Second, fully implement the network operating system and the computer operating system security features to prevent access at the operating system level. These two levels of security are not the job of the Notes Application Developer, however, and will not be tested on the exam, but they do contribute to the way in which you will begin implementing Notes security measures.

The next levels of security are Notes security measures. The following list outlines the basic options for securing the Notes environment, as also shown in Figure 11.1:

- Authentication in the Notes Organization
- Notes server access
- Database ACLs
- Form and/or View access lists
- Document access through Reader and Author fields
- Field security through encryption
- Data verification through signing
- Preventing the copying, forwarding, and printing of documents to limit access to data
- User roles that refine user's access to design elements
- Limitations in the server document for Agent access

The first two levels of security in Notes are authentication and the server control list. The server administrator—not an application developer—creates and enforces these levels. It is important, however, for you to understand the connections between methods of security in Notes.

Figure 11.1 Security Levels diagram.

Authentication is the process by which a user or server, using their ID file, verifies their identity within the Notes Organization. Each Notes Organization has a certifier (or multiple certifiers) and each ID file within the Notes Organization is stamped with this certifier. The authentication process compares the stamp on the user or server's ID file with the stamp that is part of the Notes Organization.

After using their ID to verify their identity and to gain access to the environment, the users or servers that are trying to access a database must also gain access to the server on which the database is stored. Each server's configuration document contains a *Server access list,* a list of users that are allowed (or not allowed) to access the server. The user must be allowed to gain access to the server based on the entries in this list.

Access Control List (ACL)

To determine who can gain access to a particular database and to determine which abilities the users or servers will have within that database, Notes provides an ACL for each database. The ACL determines the following three items of security:

- The users and servers that can access a database are determined by the ACL.
- The ACL defines the functions that each user, server, or group can perform.

- The ACL allows a designer to add roles for the users, servers, and groups to refine the access to the database.

To view the ACL and its options, right-click the database icon and choose Access Control, or choose File . . . Database . . . Access Control from the menus for any particular database. This displays a dialog box, as shown in Figure 11.2.

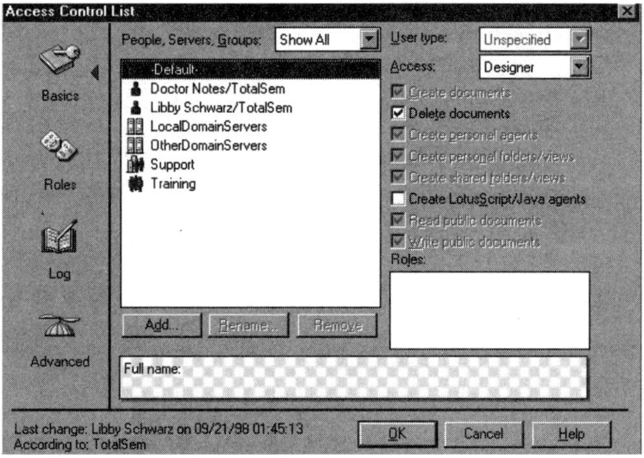

Figure 11.2 Access Control List dialog box.

The ACL has seven basic levels of security, which can then be refined through assigning additional options. These levels of security are all set on this panel of the dialog box. To assign a user or server to a level of the ACL, click the Add . . . button at the bottom of the dialog box. You then have the option to either type in the name of the user, server, or group, or to choose a name from the Public Name and Address Book. You can also remove or rename users, servers, and groups using the other buttons at the bottom of the dialog box. If a user is not listed specifically in the ACL, either individually or within a group, they are given the access level of the Default group. The Default group, which cannot be renamed or removed, is given Designer access automatically when the database is created. The user that creates the database is also added to the ACL automatically, with Manager access, when the database is created. Two other groups, LocalDomainServers and OtherDomainServers, are also added to the ACL by default. These groups are standard Notes groups that are used in all Notes environments to determine the access level of the servers that house and replicate the databases. Both of these groups have Manager access when the database is created.

NOTE
I generally recommend choosing a name, rather than typing it in, to ensure that the hierarchical name is correct. In addition, I also recommend using groups in the ACL as often as possible. This allows an administrator to change access for users by changing group membership, rather than having to open each database and change the ACL manually.

After choosing a user, group, or server, choose the appropriate user type from the drop-down list. Then choose the Access level and options from the drop-down list and the checkboxes. The following paragraphs describe each level of security and the options that are available to refine them.

No Access

The lowest level of access to a database is No Access. *No Access* prevents a user or group not only from opening a database but also from adding the database to their workspace. This level of security is often recommended for the Default access to the database so that no one can gain access to data inadvertently. The only additional options available with No Access are the ability to read public documents and to write public documents, as shown in Figure 11.3. None of the other options are selected or available.

NOTE
Public access **allows users with limited access to create or view certain documents or forms. These rights are available beginning in Release 4.5 only; previous to Release 4.5, there were no additional options available for users with No Access. Public access is necessary for creating applications in the current Notes environment, but it will not be tested on the Application Development II exam.**

Public access to documents, forms, and folders can be granted in the Form, Document, or Folder Properties InfoBoxes on the Security tab. An example of Public Access in use in Notes is the ability of users to grant other users rights to their calendars or mail files. These rights are granted by giving users with No Access the ability to read and/or write public documents.

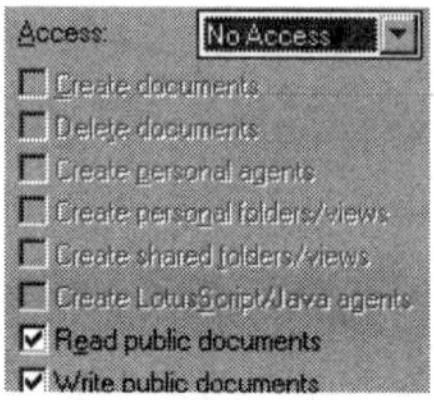

Figure 11.3 Options available with No Access to a database.

Depositor Access

The next level of access to a database is *Depositor* access. This allows a user to add the database to the workspace and open it. In addition, Depositor access users can create documents in the database, yet they cannot, however, read or edit any documents in the database; the views in the database will appear empty to a user with Depositor access.

Depositor access is useful for mail-in databases, surveys, suggestion boxes, or evaluations. The users can create an evaluation for an instructor, for example, but will not be able to see or change any of the evaluations. The Create documents option is checked by default and cannot be changed. No other options (except for the read and write public documents options in Release 4.5 and above) are available, as shown in Figure 11.4.

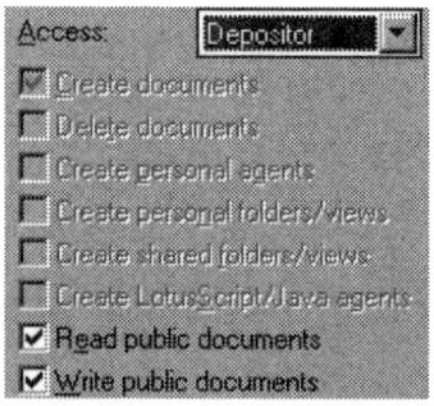

Figure 11.4 Options available with Depositor access to a database.

Reader Access

You can also assign *Reader* access to a database. Users with Reader access to a database can view documents in the database, but cannot create or edit documents in the database. This type of access is useful for informational databases, such as a corporate directory or a documentation database. The additional options for users with Reader access are Create personal agents, Create personal folders/

views, and Create LotusScript agents, as shown in Figure 11.5. The other options are all unchecked and unchangeable with the exception of Read public documents in Release 4.5 and above, which is checked and unchangeable.

Figure 11.5 Options available with Reader access to a database.

Author Access

The next level of access to a database is *Author* access. Users with Author access have the ability to read documents in the database, create documents in the database, and edit the documents they create. Author access is the most common level of access to databases, as it helps to prevent *replication and Save conflicts*. The additional options available for users with Author access are Create documents, Delete documents, Create personal agents, Create personal folders/views, and Create LotusScript agents, as shown in Figure 11.6. In Release 4.5 or above, the Read public documents option is checked and unavailable for change and the Write public documents option is available for change.

Figure 11.6 Options available with Author access to a database.

NOTE

A replication or save conflict is an error document created when multiple users edit the same document. Options to help prevent replication and save conflicts have been mentioned multiple times in the book.

Editor Access

Users with *Editor* access to a database start with the same level of access as users with Author access: the ability to read, create, and edit their own work. Editor access also grants the ability to edit all other documents in the database. It is usually recommended to give only one or two users this level of access to a database to prevent *replication and Save conflicts*. The Create documents option is checked by default, as shown in Figure 11.7, and is not available for change. In Release 4.5, this is also true of both the Read public documents and Write public documents options. All other options are available for modification for users with Editor access. This is the first level of access that can grant the ability to create shared folders and views.

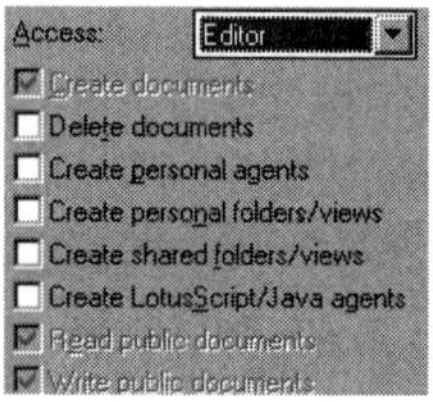

Figure 11.7 Options available with Editor access to a database.

Designer Access

Designer access gives a user all the rights allowed by Editor access including reading, creating, and editing all documents in the database. Designers have the additional ability to create and modify all design elements in the database including forms, views, navigators, and others. Designers have the ability to create and modify full text indexes, About and Using this database documents, and database icons. The Create documents, Create personal agents, Create personal folders/views, and Create shared folders/views options are granted to users with Designer access by default and cannot be changed. In Release 4.5, this is also true of both the Read public documents and Write public documents options. The options for Delete documents and Create LotusScript agents are available to be modified, as shown in Figure 11.8.

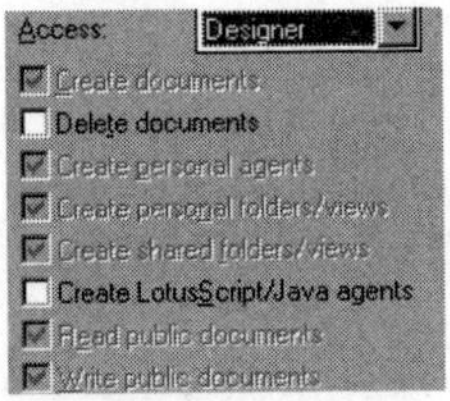

Figure 11.8 Options with Designer access to a database.

Manager Access

Users with *Manager* access are granted all the abilities of users with Designer level access as a base. In addition, this is the only access level in which users are given the ability to change the ACL of the database. Managers can also encrypt a database, delete a database, and set the replication settings for a database. All options are granted by default to Managers and cannot be changed, with the exception of the Delete documents option, as shown in Figure 11.9. There should usually be only one user or group with Manager level access, and one server with Manager level access, to prevent problems with security and replication.

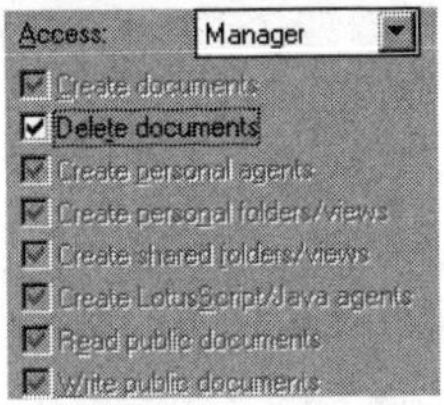

Figure 11.9 Options with Manager access to a database.

The application developer and the Notes system administrator should work together to create and maintain the ACL for each database. Some aspects of security can affect how the application functions, which is the responsibility of the application developer or the database manager. On the other hand, some aspects of the ACL can affect the security of the Notes environment as a whole and can affect replication of the databases throughout the environment; both of which are the responsibility of the Notes system administrator. In addition, in a secure Notes environment, the database designer may not be given Manager level access to the database after it is placed on the server, although the designer had this access when he or she created the database.

View and Form Access Lists

After a user is given access to a database through the ACL, the user then has to gain access to the forms and views in the database. These forms and views have a Security tab on their Properties Info-Boxes that allows the designer to refine the access given to users. First, we will examine the View (and Folder) access lists. Then we will discuss Form (and Document) access lists.

View Access Lists

In the View access list, the developer can determine which users can use the view. To see the View access list, open the View Properties InfoBox and click the Security tab (the key icon). All users with Reader access and above to a database are given access to the Views by default, as shown in Figure 11.10. If you want to create a view that only certain users can access, remove the check mark from the All readers and above option. You then have the option to add checkmarks next to the users, servers, and groups that should be able to gain access to the view. The blue Person button to the left of the dialog box can be used to choose names from the Public Name and Address book. The users must, however, be given access to the database through the database ACL to use the view. If you exclude users from the View access list, they will no longer be able to see the view on the View menu.

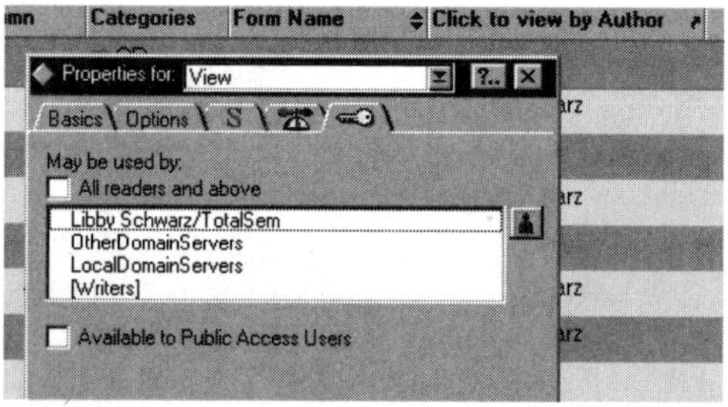

Figure 11.10 View Security options.

The same access list option is available for shared folders in the database. The designer either can accept the default, which allows all users with Reader access and above to view the contents of the folder, or can choose the users, servers, and groups that are given this access. The folder security contains an additional option, however, as shown in Figure 11.11. The additional option for folders allows the designer to determine which users can update the contents of the folder. By default, all authors and above are given the ability to add and remove documents to and from a folder. The designer is also given the ability to refine this access by removing the ability to change the contents of the folder from certain users' privileges.

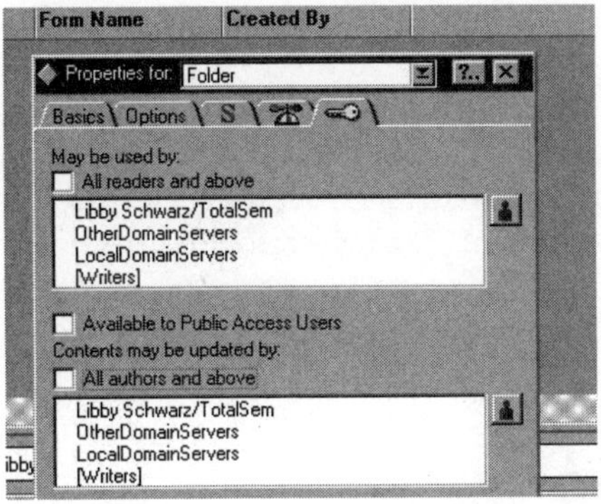

Figure 11.11 Folder Security options.

In both cases, the View and Folder access lists should not be considered true security measures. A user with at least Reader access to the database could create a private view or folder to display the data. The View and Folder access lists should be used in combination with other security measures for the database.

Form Access Lists

Access to forms can be refined by the designer with a similar method. To view the Form access list, open the Form Properties InfoBox to the Security tab, as shown in Figure 11.12. The first option on this tab is *Default read access for documents created with this form.*

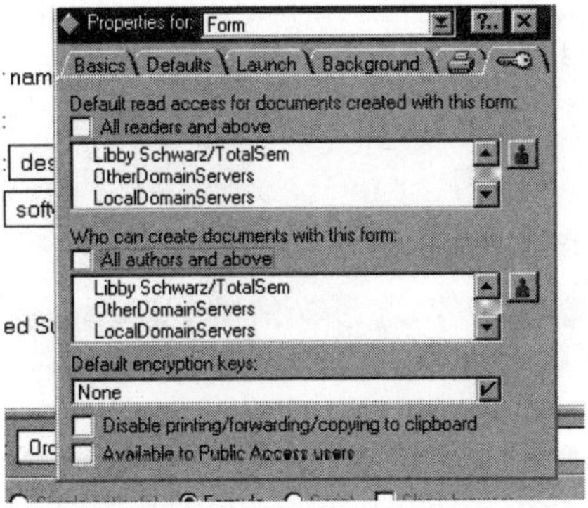

Figure 11.12 Form Security options.

All users with Reader access and above are given this ability by default. When you refine this access by removing this ability from all users with Reader access and giving it only to certain users, a reserved field is added to the document. This field ($Readers) can be viewed by opening the Document properties created by this form and examining the Fields tab, as shown in Figure 11.13. If there are other Readers fields on the form, Notes combines all users listed in the Readers fields and gives them all default Reader access to the documents created with the form. Note that a user must have access to the database through the database ACL to view documents in the database. If a Read access list exists, however, a user must be on that list to read documents created by the form, regardless of ACL level.

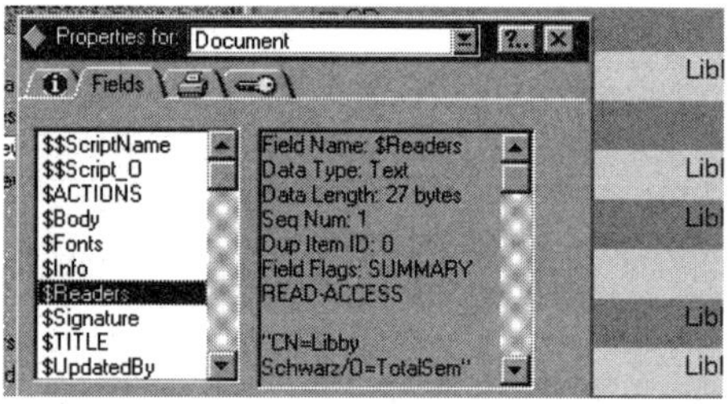

Figure 11.13 $Readers field shown in Document Properties InfoBox.

The second option on the Form properties Security tab is the list of the users that can create documents with the form. By default, all users with Author access and above are given this ability. Remember that this access list is only able to refine the ACL, not change it. This means that a user with Reader access in the ACL will not be able to create a document with the form, even if you have listed them in the Form access create list. Similarly, if a user has Editor access or above and is able to read the documents created with this form (he is on the Read access list), that user will be able to create documents with this form, regardless of the Create access list. If a user is removed from the Create access list and is not covered by the exception previously mentioned, that user will not see the form listed on the Create menu.

The Access list options created for forms provide the defaults for the documents created with those forms. Each document, however, can refine this default list further. The document security options, available from the Document Properties Security tab, allow the author or editor of the document to determine who can read the document. These options are shown in Figure 11.14.

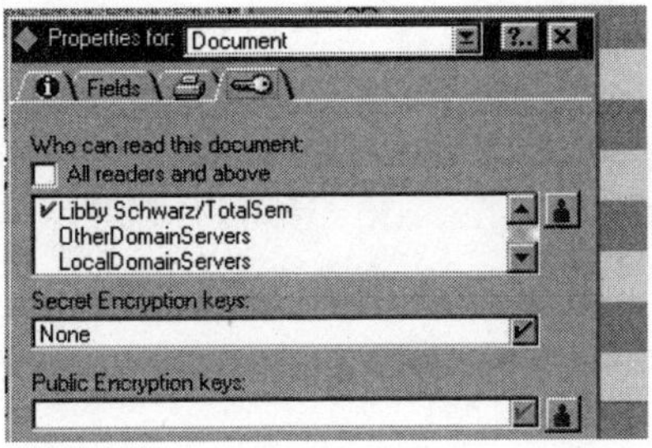

Figure 11.14 Document security options.

Reader and Author Fields

Although default access to documents is granted through the settings on the form and in the ACL, as described previously, a designer can further limit access to documents through the use of Readers and Authors fields. Readers and Authors fields can be either editable or computed and must resolve to a list of names. As in the View and Form access lists described previously, Readers and Authors fields refine and limit the abilities given in the ACL; they cannot override the ACL.

Readers Fields

A Readers field on a document limits who can read that document. To create a Readers field on a document, create a field of the Readers data type, either editable or computed, that resolves to a text list of names. If the field is editable, the designer gives the authors or editors of the document the ability to choose the users that have read access to the document, as shown in Figure 11.15.

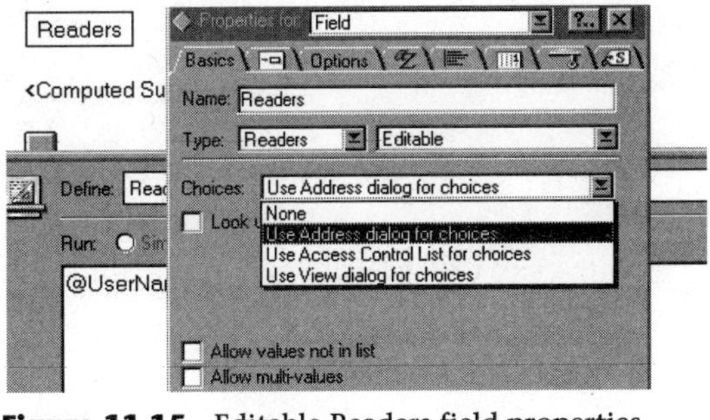

Figure 11.15 Editable Readers field properties.

The designer has the following options for allowing the author or editor to choose users for the Readers field:

- *Use Address dialog for choices* allows the author or editor to select names from the Personal or Public Name and Address book. The dialog box shown in Figure 11.16 will be displayed.

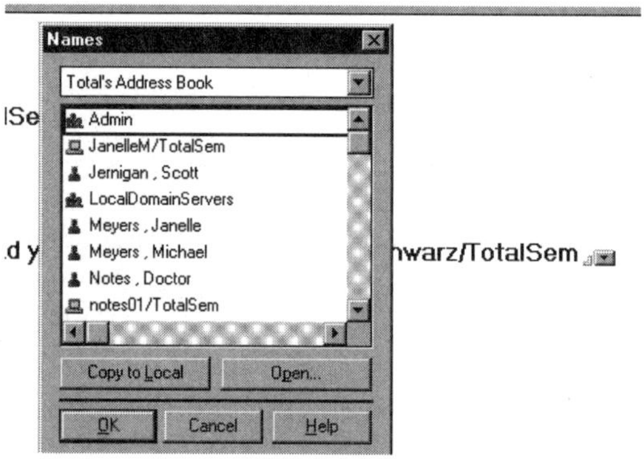

Figure 11.16 Address dialog for choices.

- *Use Access Control List for choices* allows the author or editor to use names from the ACL to determine the users that can read the document. Users will see the dialog box shown in Figure 11.17.

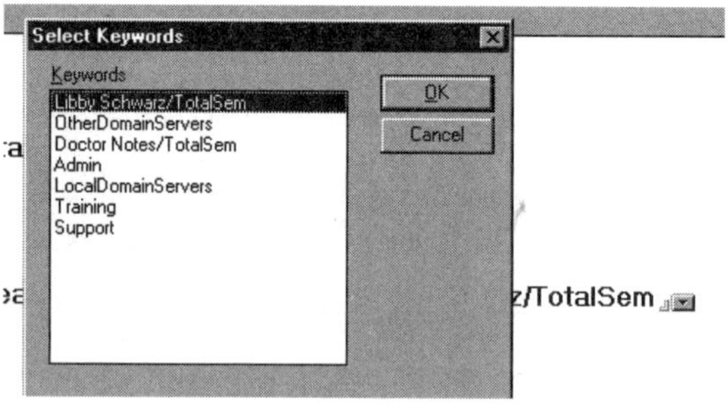

Figure 11.17 Access Control list dialog box for choices.

- *Use View dialog for choices* allows the designer to use data from a column in a view as the choices for the Readers field. The user will see the dialog box shown in Figure 11.18.

Figure 11.18 View dialog box for choices.

When you make the Readers field editable, you may want to add the function @Author or your own name as a default value to ensure that the document will have at least one reader. In addition, you may want to use an input validation formula to ensure that this field gets completed.

If the Readers field is computed, create a formula that resolves to a list of user, group, or server names. The names in the formula should be enclosed in quotation marks and concatenated using colons. The following formula, for example, would allow the user Libby Schwarz, the group Instructors, and the server ServerZ to read this document:

```
"Libby Schwarz" : "Instructors" : "ServerZ"
```

You may also want to use a formula that chooses readers based on values of other fields in the document or roles assigned in the database.

When using Readers fields, it is important to recognize that they interact with other security aspects of the database to provide the complete security picture. Assuming a user has at least Reader access to the database, the following rules determine who can read the documents:

- If the document contains a Readers field, and the user is listed in the Readers field, the user can read the document.
- If the document contains a Readers field, and the user is not listed in the Readers field, the user cannot read the document, regardless of the user's level of access in the ACL.
- If a document contains a blank Readers field, all users with reader access and above in the ACL can read the document.

- If a document contains multiple Readers fields, all users listed in all the fields can read the document.
- If a document uses a Read access list, as described in the preceding sections, all users listed in the Read access list and in any of the Readers fields can read the document.
- If a document contains a Readers field and an Authors field, users that are in the Authors field can read the document, even if they are not listed in the Readers field. Authors fields are discussed later in this section.

Authors Fields

An Authors field on a document limits who can edit it. This field can give Editor access for specific documents to users that have Author access to the database but did not create the document. Again, an Authors field does not override the ACL. Only users that have access to the database at the Author level are affected by an Authors field. Essentially, this type of field gives Editor access to a document without giving Editor access to the database.

To create an Authors field on a document, create a field of the Authors data type, either editable or computed, that resolves to a text list of names. If the field is editable, the designer gives the authors or editors of the document the ability to choose the users that can edit the document, as shown in Figure 11.19.

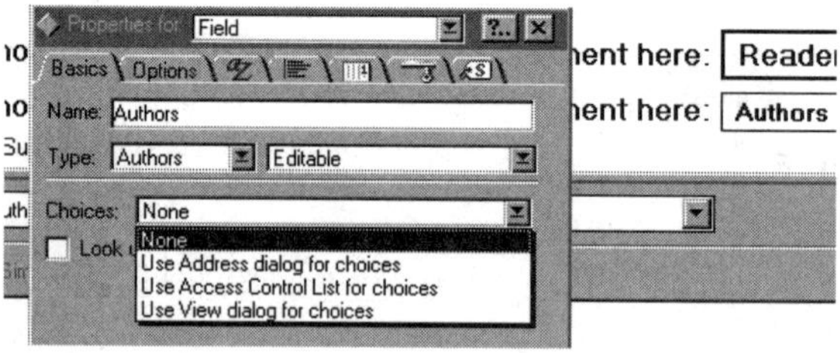

Figure 11.19 Editable Authors field properties.

The designer has the following options for allowing the author or editor to choose users for the Authors field:

- *Use Address dialog for choices* allows the author or editor to select names from the Personal or Public Name and Address book. The dialog box, as in Figure 11.20, will be displayed.

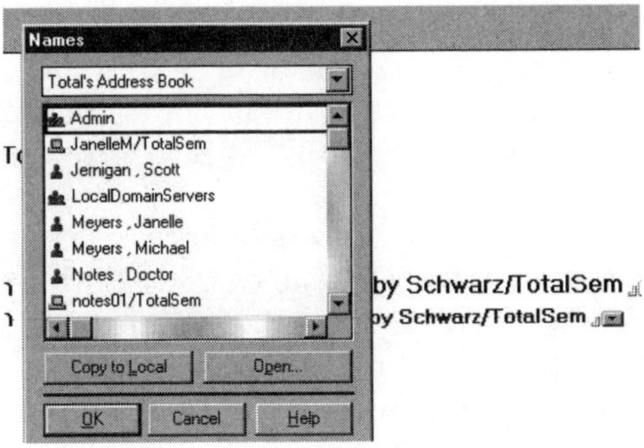

Figure 11.20 Address dialog box for choices.

- *Use Access Control List for choices* allows the author or editor to use names from the ACL to determine the users that can edit the document. Users will see the dialog box shown in Figure 11.21.

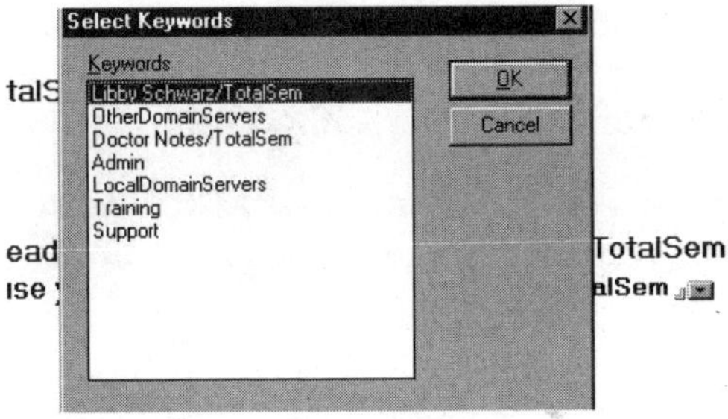

Figure 11.21 Access Control list dialog box for choices.

- *Use View dialog for choices* allows the designer to use data from a column in a view as the choices for the Authors field. The user will see the dialog box shown in Figure 11.22.

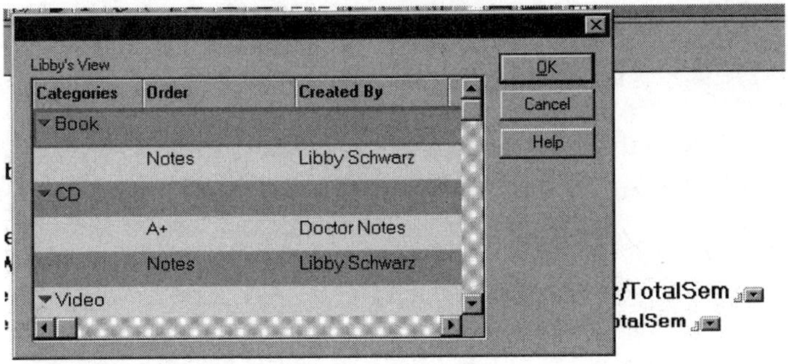

Figure 11.22 View dialog box for choices.

When you make the Authors field editable, you may want to add the function @Author or your own name as a default value to ensure that the document will have at least one editor. In addition, you may want to use an input validation formula to ensure that this field gets completed.

If the Authors field is computed, create a formula that resolves to a list of users, groups, or server names. The names in the formula should be enclosed in quotation marks and concatenated using colons. The following formula, for example, would allow the user Libby Schwarz, the group Instructors, and the server ServerZ to edit this document:

"Libby Schwarz" : "Instructors" : "ServerZ"

You may also want to use a formula that chooses editors based on values of other fields in the document or roles assigned in the database.

When using Authors fields, like Readers fields, it is important to recognize that they interact with other security aspects of the database to provide the complete security picture. The following rules determine who can edit documents when they contain Authors fields:

- Users with Reader access or below in the database ACL will not be affected by an Authors field. They *will not* be able to edit documents in the database.
- Users with Editor access and above in the database ACL will not be affected by an Authors field. They *will* be able to edit documents in the database.
- In a document with multiple Authors fields, a user listed in any one of the fields can edit the document.
- In a document with an Authors field, if the field is blank, only users with Editor access or above in the database ACL will be able to edit the document. To ensure that the creator of a document can edit the documents they create, make part of the value for an Authors field an @Author function.

The Authors field only contains a list of potential editors for a document. To see the users that have actually edited the document, look at the reserved $UpdatedBy field. The $UpdatedBy field can be viewed by looking at the Fields tab on the Document Properties InfoBox or by clicking an Authors field in a document. When the user clicks an Authors field, only the users that have actually edited the document are displayed, using the $UpdatedBy reserved field information.

NOTE
For both Readers and Authors fields, either the database must be located on a server or the Enforce Consistent ACL option must be enabled to see the results of these fields. This is true for most database security options in Notes.

Encryption

After gaining access to the server, database, form or view, and document, an unauthorized user can still be denied access to data at the field level. Security at the field level is applied through *encryption,* which means taking the raw data stored in a field and encoding it so that only a user that has the correct key can decode and read the data. Field-level encryption allows a document's author or editor to apply one or more encryption keys to a document. When the encryption key(s) is applied, any data in encryption-enabled fields on the document is secured through encryption.

When adding encryption to the databases you design, you will have to complete the following four steps:

1. Enable the fields in the database for encryption.
2. Create the secret encryption key or keys.
3. Apply the key in one of three ways.
4. Distribute the key or keys to the users that will need access to the encrypted information.

Enabling Fields for Encryption

Before encryption will encode information on the documents that you create, you must first create fields that can be encrypted. To apply this property to fields on the forms you have created, open the Field Properties InfoBox for the field to the Options tab, as shown in Figure 11.23. At the bottom of the dialog box is a drop-down list for security options for that field. Choose Enable encryption for this field from the drop-down list. This field now has the ability to be encrypted.

The field that you have made encryption-enabled can be of any data type. When you encrypt Rich Text fields, however, the following additional rules apply:

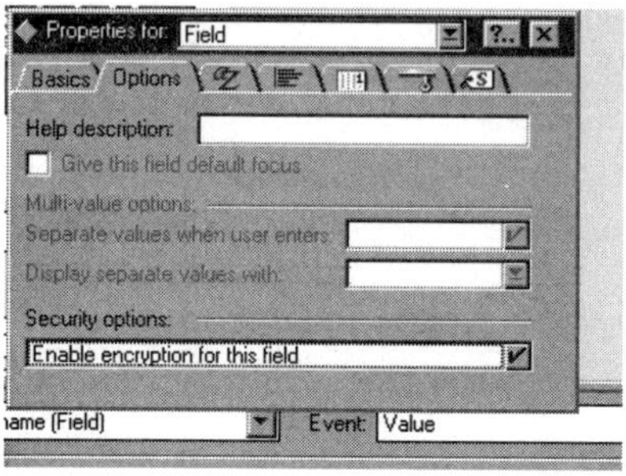

Figure 11.23 Enable encryption for a field.

- Any OLE object in an encrypted Rich Text field will only be visible to the users that have the correct encryption key.
- Any attachments in an encrypted Rich Text field are visible to any users that can read the document, even if they do not have the correct encryption key.

After you have given the field the capacity to be encrypted, you must create encryption keys to do the encrypting.

Create an Encryption Key

Before documents can be encrypted, the correct encryption keys must be created. Encryption keys are created, stored, and used from the user ID file. To create an encryption key or to view the encryption keys stored in your user ID file, open your user ID by choosing File . . . Tools . . . User ID from the menus. When the User ID dialog box appears, choose the Encryption panel, as shown in Figure 11.24. From this panel, you can create, delete, import, export, and mail encryption keys. Note that you have the ability to delete an encryption key in this panel. If you choose to delete the key, it is permanently deleted and is unrecoverable unless it is also stored in another user's ID file or has been exported to a file and stored elsewhere.

To create the new key, click the New . . . button. The Add Encryption Key dialog box, shown in Figure 11.25, allows you to do the following:

- Name the Encryption key.
- Decide whether the Encryption Key will be used in North America only or internationally.

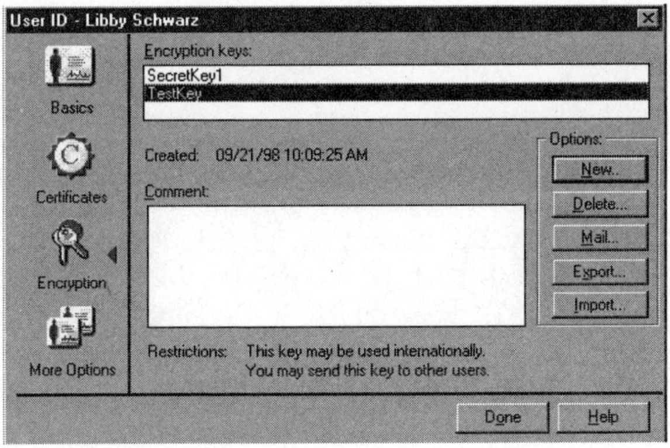

Figure 11.24 Notes ID encryption options.

- Comment on the Encryption key to ensure that not only do you remember its purpose, but also ensure that the users that you send it to remember the purpose. This will help prevent users from deleting the encryption keys accidentally.

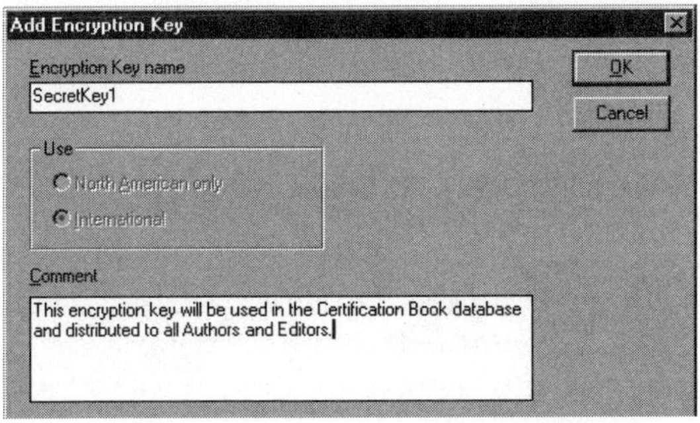

Figure 11.25 Add encryption key dialog box.

After you create the encryption key or keys, you will need to apply those keys to the database and distribute those keys to other users.

Apply the Encryption Key

After you have created an encryption key, you have three options for applying that key to a form or a document in the database:

- *Forced encryption* is encryption that is applied to the form by the designer. To apply the encryption key in this manner, open the Form Properties InfoBox to the Security tab (the key icon), as shown in Figure 11.26. Choose a key or keys from the Default Encryption Keys drop-down list. This key is then applied by default to this form. Any fields that have been made encryption-enabled will be encrypted by this key or keys. Only users that have the key or keys in their user ID files will be able to read the data in those fields.

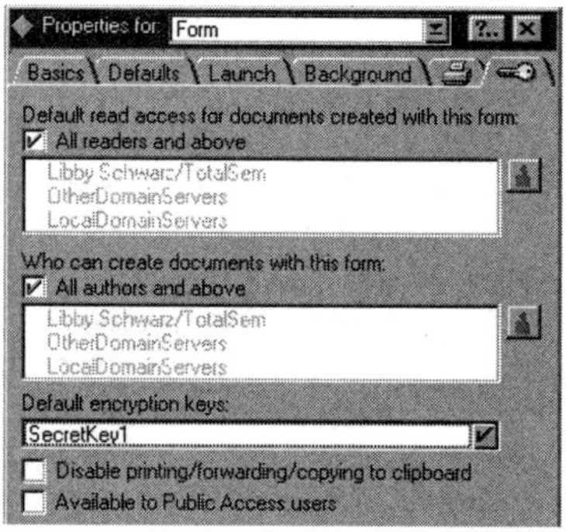

Figure 11.26 Form Properties InfoBox, Forced Encryption.

- *Optional encryption* is enabled by the designer, but it is either chosen by the user through a keyword field or determined through a formula. To apply optional encryption to a form, create a field with the reserved field name SecretEncryptionKeys. This reserved field should be either a keyword field or a text field. It can be editable or computed, but must resolve to the name of a Secret Encryption key.

  To use a keyword type of field, make the synonyms of the keyword choices the names of the encryption keys. Figure 11.27 shows three keyword choices. If a user chooses the keyword Secret, the encryption key called Key1 is applied. If a user chooses the keyword Confidential, the encryption key called Key2 is applied. If a user chooses the keyword Open, no encryption is applied to the document.

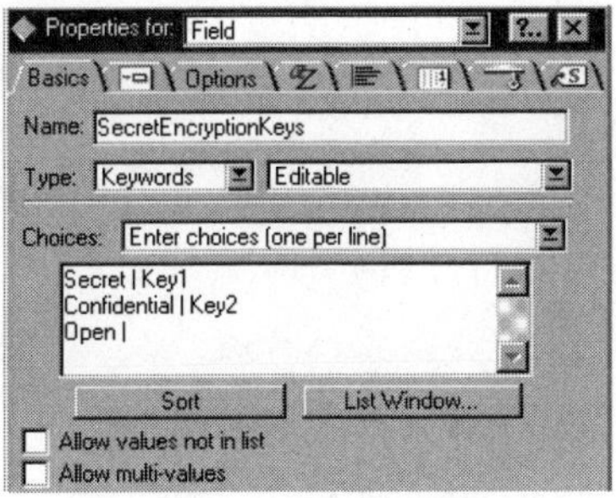

Figure 11.27 Keyword Encryption.

To use a text field, create a field (still using the reserved fieldname SecretEncryptionKeys) that has a formula that will resolve to the name of an encryption key. If you want to base encryption on the status of a document, for example, you could write a formula like this one:

```
@If (Status = "Approved"; "Key1"; Status = "Pending"; "Key2";
"")
```

This formula checks the field status and applies Key1, Key2, or no encryption based on the value of the field.

NOTE
Note that encryption key names are case-sensitive.

- *Manual encryption* is applied by the authors and editors specifically to a document. The designer still creates fields that are encryption-enabled and, when the user creates or edits the document, the user can open the Document Properties InfoBox to the Security tab, as shown in Figure 11.28. The user can then select an encryption key from the Encryption keys drop-down list. The user can also choose any of the secret encryption keys that are stored in their user ID files, but would have to distribute the secret key to other users that need access to the data after encrypting the document.

Figure 11.28 Applying manual encryption using Document Properties.

Here are a few final points to remember about applying encryption keys:

- A designer or user can apply multiple encryption keys to all of the fields in the forms and documents. If they do, a user will need all the encryption keys to view any of the data.
- If an encryption key is selected, that encryption key will encrypt all encryption-enabled fields on the document.
- Users with Author access can apply or remove encryption from any document that they have created once they have the appropriate encryption key.
- Users with Editor access or higher can apply or remove encryption from any document in the database once they have the appropriate encryption key.

Distributing the Keys

After the keys have been created and applied, the keys must be distributed to the appropriate users. Without the encryption key stored in their user ID file, the users will not be able to access the data in the encrypted documents. To distribute the encryption keys that you create, you can use one of the following two methods:

- *Use Notes Mail to distribute the encryption keys.* After creating the encryption key, as described earlier in this section, you can choose to distribute the key using Notes mail. In the User ID dialog box, select the key that you want to distribute and click the Mail button. The Mail Address Encryption Key dialog box, shown in Figure 11.29, is displayed. The encryption key will be distributed to any users listed in the To: or CC: fields. You can use the Address button to choose names from the Name and Address book, if necessary. The Subject: field describes the key and tells the recipient how to accept the key. Before sending the key, you will be prompted to decide whether the user should have the right to distribute this key to other users. If you allow this, the users to whom you send the key can then send it to other users. This can limit the security effectiveness.

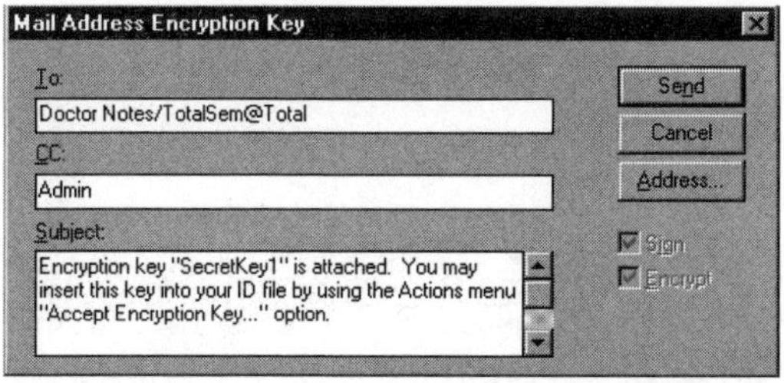

Figure 11.29 Mail the Encryption key.

- *Use a file to distribute the encryption keys.* You also have the option of distributing the key by exporting it to a file and then giving this file to other users. To export the key to a file, select the Export button in the User ID dialog box. The User ID Encryption Key Export dialog box is displayed, as shown in Figure 11.30. You are prompted to secure the encryption key with a password. If you choose not to secure the file, click the No Password button.

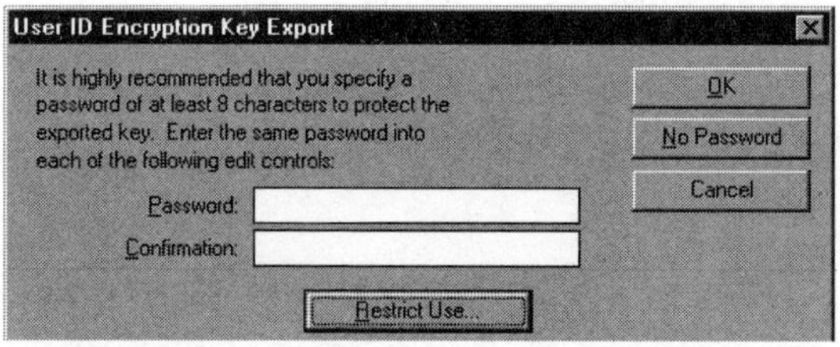

Figure 11.30 User ID Encryption Key export dialog box.

You can restrict the use of the encryption key to one user and choose whether the encryption key stored in the file can be forwarded to other users by clicking the Restrict Use button, as shown in Figure 11.31.

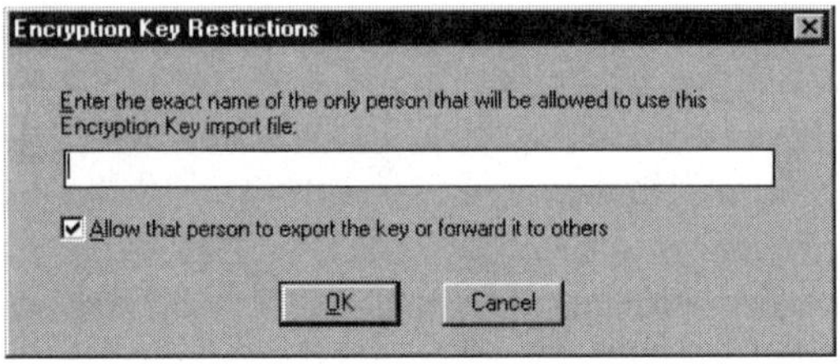

Figure 11.31 Encryption Key restrictions.

Finally, you must give the encryption key file a name and choose the storage location. The file will have a .KEY extension. You can then distribute the file to other users.

Accepting an Encryption Key

When the users receive encryption keys, either through Notes mail or in a file, they must add those keys to their user ID files. To accept a key that has been mailed, the user can choose Actions . . . Accept Encryption Key from the menus while they are in the mail message with the encryption key. The Accept Encryption key dialog box, shown in Figure 11.32, is displayed to the user. The user then clicks the Accept button and the encryption key is stored in the user ID file.

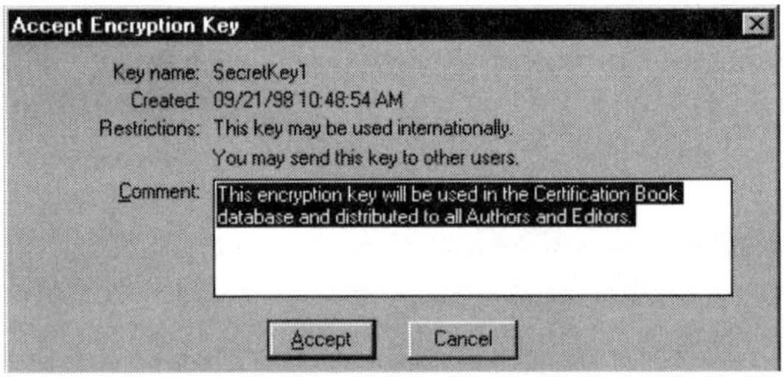

Figure 11.32 Accept Encryption Key dialog box.

If the encryption key is distributed through a .KEY file, the user must import the file into their user ID. To do so, open the ID file dialog box and choose the Import button. Choose the encryption key file and click Open. The Accept Encryption key dialog box is displayed and the user can click Accept to import the file into their user ID.

Mail Encryption

Some encryption, such as mail encryption, is done through a combination of the Public and Private keys stored in the Public Name and Address book and in the user ID file. The public key, which is created for each user when that user is registered, is stored in the user's Person document in the Public address book. Part of a public key is shown in Figure 11.33. The public key is used to encrypt the mail message.

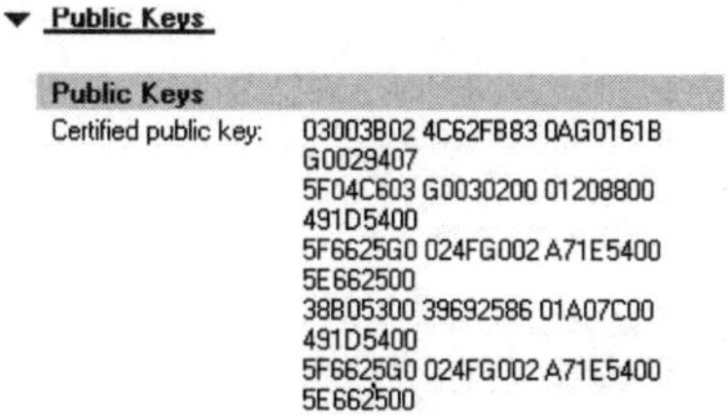

Figure 11.33 Public Key stored in a Person document.

The private key is stored in the user ID file and is used to decrypt the mail message. Only the user that has the appropriate user ID file can decrypt the mail message. Mail encryption is used on documents that are being sent to another user's mail database. Mail encryption can be accomplished in one of three ways:

- *Encrypt incoming mail.* When a user chooses to encrypt incoming mail, the messages are encrypted at the server. The mail messages cannot be read by administrators or by the server. This option can be set in the user's Person document in the Public Name and Address Book, as shown in Figure 11.34.

Figure 11.34 Encrypt incoming mail option in the Person document.

- *Encrypt outgoing mail.* When a user chooses to encrypt outgoing mail, the messages are encrypted as you send them to other users.

The message cannot be accessed by administrators or servers at any point in the transfer or delivery process. This option can be set in the multiple ways:

- By going to File . . . Tools . . . User Preferences . . . Mail . . . Encrypt Sent Mail.
- NOTES.INI variable SecureMail = 1.
- Choose Encrypt in the Delivery Options dialog box when sending mail messages, shown in Figure 11.35.
- Designers can force mailed documents to be encrypted by using a field called Encrypt and setting the value equal to 1.
- Designers can force mailed documents to be encrypted by using the [Encrypt] flag with the @MailSend function.

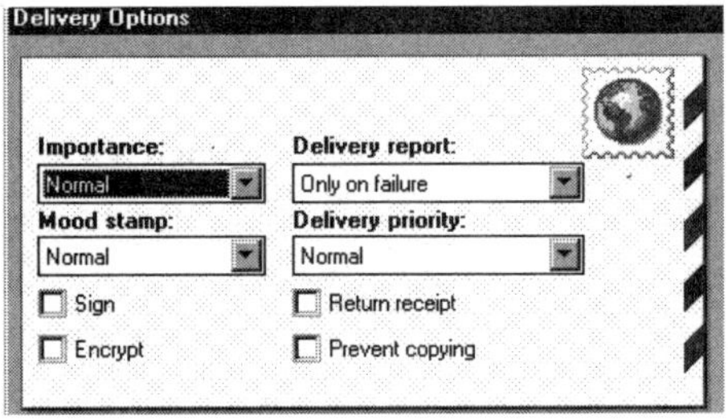

Figure 11.35 Encrypt in the Delivery Options dialog box.

- *Encrypt saved mail.* Encrypting saved mail prevents messages stored in your server or local mail database from being accessed by unauthorized users. Many laptop users, especially, use this type of encryption to protect their data in case of losing their laptop. To set this option, choose File . . . Tools . . . User Preferences . . . Mail . . . Encrypt Sent Mail. The Encryption options available as a user preference are shown in Figure 11.36.

If a document has secure fields or data and is being mailed, use both types of encryption: mail encryption and document encryption. This requires the designer both to add a SecretEncryptionKeys field that resolves to the name of an encryption field and to add an Encryption field with a value of 1. Both types of encryption must be applied because the mailed document would be sent before the document would be saved and encrypted for storage in the local database.

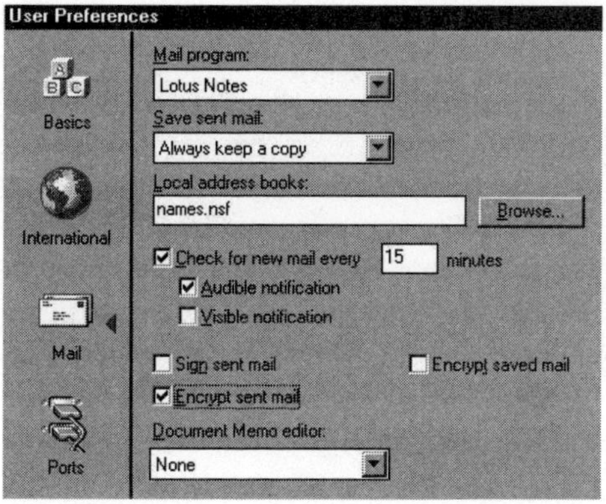

Figure 11.36 User Preferences Encryption options.

Additional Security Options

Some additional security options exist that you may want to apply to your database. These security options include signing; preventing documents from being forwarded, printed, or copied; and adding user roles to the database. We have already discussed other security options, such as agent security and controlled access sections.

Signing

Signing is a process that places a verification stamp on a document that allows the reader to confirm that the signer was the last person to modify the document. The signature is created using the user's Private key (discussed previously) in combination with the data in the field. The signatures can be attached either to sections or to fields within documents.

Any field on any form can be enabled for signing. To enable a field for signing, open the form in design mode and access the field's Properties InfoBox. On the Options tab, shown in Figure 11.37, choose the Sign if mailed or Saved option from the Security Options drop-down list. Although all field types can be signed, it is usually best not to sign the reserved mail fields (SendTo or CopyTo), as these can be updated during the mailing process and can be changed and resigned.

When you sign a single field on a form, the entire form is signed. To add multiple signatures to a form, use sections. To sign a section, create a field in each section that is signing-enabled. When the document is saved, the section is signed. This is different from signing outside a section; the document is only signed when it is mailed if the signing enabled field is not in a section. Signing sections is often used with controlled access sections. Any time a possible editor of the section saves a document, the signature on the section is updated. It may be most useful to create sections that have only one or two possible editors in this situation.

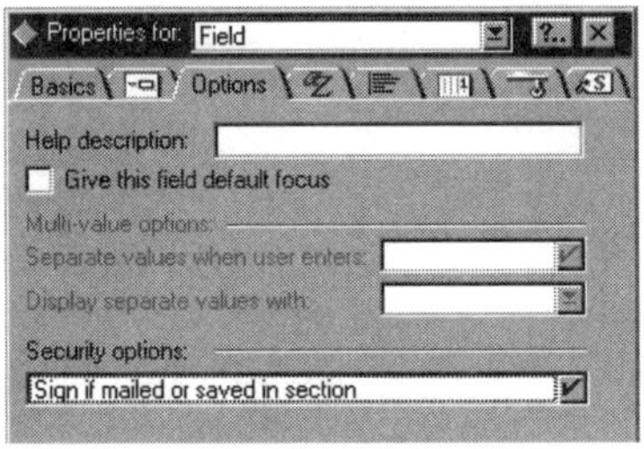

Figure 11.37 Sign a field.

Prevent Printing, Forwarding, and Copying

It is possible to prevent users of the documents from printing, forwarding, or copying the data on a document that was created with a specific form. To enable this option for a form, open the Properties InfoBox for the form to the Security tab, as shown in Figure 11.38. Enable the checkbox that controls the Disable printing/forwarding/copying to clipboard option. This prevents a user from using the Forward action on the Actions menu. It also prevents a user from printing documents created with this form. Finally, it prevents a user from copying the document or the data on the document to the clipboard. It is not, however, a true security measure. A user could use the Print Screen option or a Screen Capture utility to place the data in a graphic or print the data. Instead, this option is most useful for preventing accidental sharing of sensitive information.

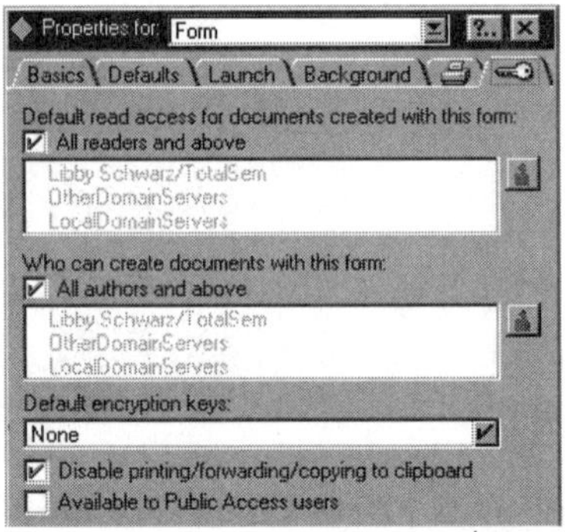

Figure 11.38 Prevent users from printing, forwarding, or copying a form.

Roles

A designer also has the ability to refine security for an application further by creating and using user roles. User roles are created in the ACL dialog box and allow the designer to gather users, servers, and groups that should have similar rights and abilities within a specific database. Roles provide the following advantages to the designer or manager of a database:

- Roles help provide centralized security. Instead of changing user names in every Form or View access list, Readers or Authors field, and section within a database, designers and database managers change the users within a particular role or change the access of the role.
- Roles are listed in the ACL of a database, ensuring that when changes are made to the security of the database, the changes take affect in the design elements. If the security is all done at the level of the design elements, many database managers and designers might neglect to update the security correctly.

To create a role, open the database Access Control dialog box to the Roles panel, as shown in Figure 11.39. To add a new role to the database, click Add and type in the name of the role that you want to add. The name can be up to 15 characters. You can also rename or remove user roles from the database's ACL from this panel on the dialog box. The database manager can add up to 75 roles to the database.

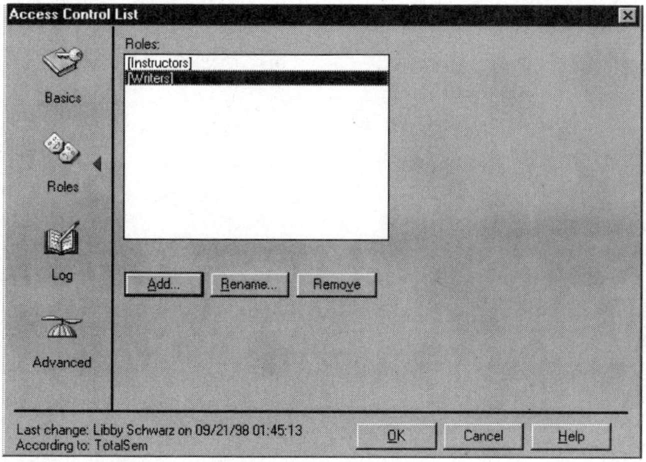

Figure 11.39 Create a User Role.

After creating the user roles, you must apply those roles to the users, servers, and groups listed in the database ACL. To apply a role to a user, open the Basics panel on the ACL dialog box, as shown in Figure 11.40. Select the user and choose the role or roles that the user should have from the Roles list on the right-hand side of the dialog box. This will associate this user role with the selected user, server, or group.

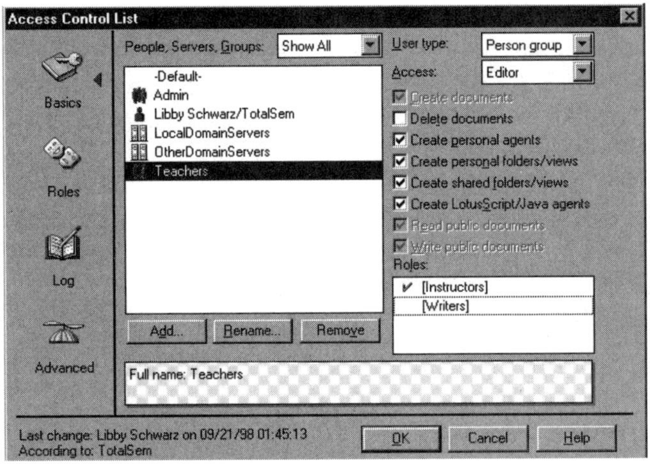

Figure 11.40 Apply the User Role

After creating and applying user roles, the user roles must be given function within the database. User roles can be used in the View and Form access lists, in Readers or Authors fields, and in Section access lists. They are indicated in a formula by enclosing the name of the role as a text string within brackets, such as [Instructors]. In addition, the function @UserRoles can be used to control access. The @UserRoles function returns a text string list of all the roles assigned to the current user. This can be used to determine the value of a field, the access to a section, the result of a Hide-When formula, or the computed subform that is displayed to a user. If you only want users with the [Instructors] role to see the answers to questions in a database used for testing, for example, you might create Hide-When formulas that use the following function:

@IsNotMember ("[Instructors]";@UserRoles)

This function examines the roles of the current user to see if [Instructors] is included. If the user is not a member of this group, the text controlled by this Hide-When function is hidden.

Sample Database

You have placed the database on the server and it is almost ready for use. Before you turn your users loose on it, you should create and enforce some security measures. For example, the Answers form and the Overdue view need to be limited in use. You do not want your end users answering the questions of their coworkers. You also do not want your users to see a list of problems that your Help Desk is late in answering. To prevent these issues from occurring, use Form and View access lists to determine which users can use these design elements. Training and Support are both groups in the ACL of the database that should be given the capability to use these design elements. To make it easier to assign users to these lists, we can create a HelpDesk role that will be given Create access to the Answers form and read access to the Overdue view.

Create the role using the Roles panel on the ACL, as shown in Figure 11.41. Then apply the role, as shown in Figure 11.42, to the Support and Training groups in the ACL. You will also want to give yourself, your system administrator, and the necessary servers any necessary roles.

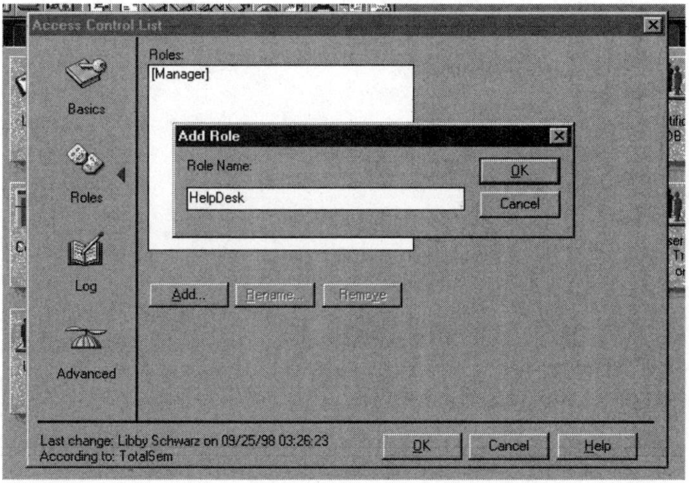

Figure 11.41 Create a HelpDesk role for the database.

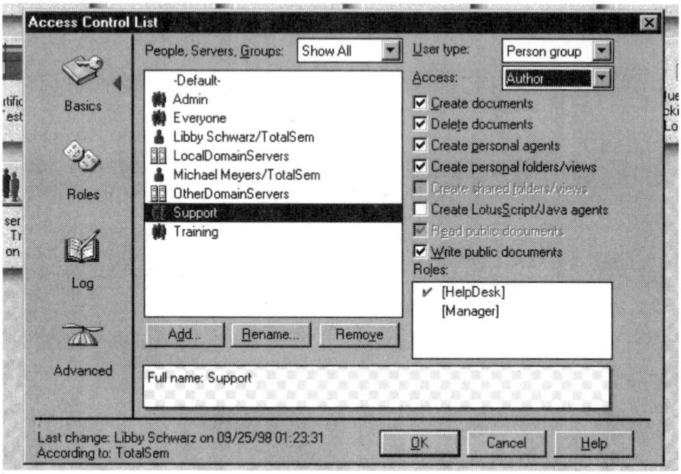

Figure 11.42 Applying the HelpDesk role in the database.

Now that you have created the HelpDesk role, you can use that role to assign the ability to create documents with the Answers form. Note that you do not want to limit the ability to read the Answers form, as all your users will be reading documents created with it! The Create control list for the Answers form is shown in Figure 11.43. Figure 11.44 shows the View control list for the Overdue view.

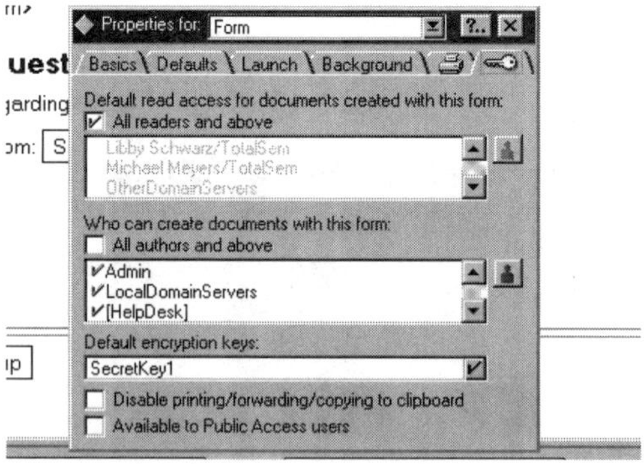

Figure 11.43 Answers form Create control list.

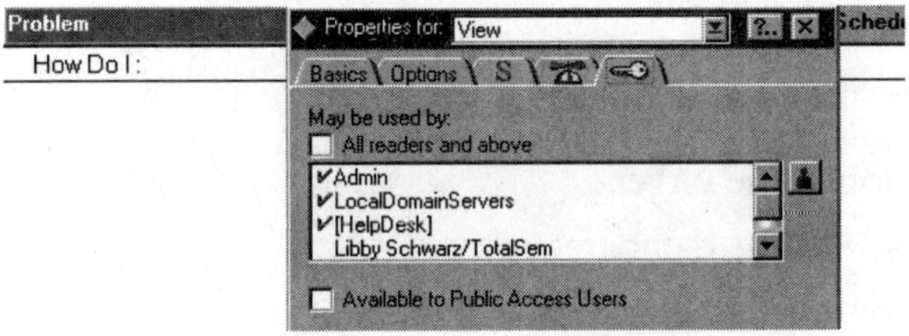

Figure 11.44 Overdue view access control list.

Some of the managers have suggested that it might be nice for the users to be able to verify that the answer to their questions is coming from a real member of the help desk group. To do this, you might decide to sign the document. The Notes field in the Internal section of the Answer document is a perfect field to make signing-enabled. To make the field signing-enabled, use the Field Properties InfoBox, as shown in Figure 11.45.

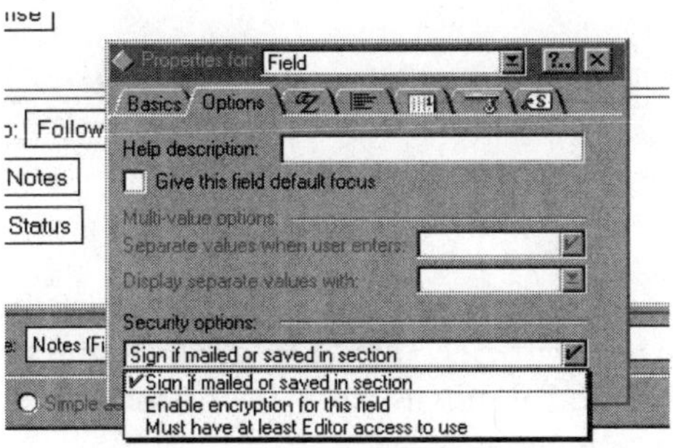

Figure 11.45 Make the Notes field signing-enabled.

Review Questions

1. The ACL determines what aspects of security:

   a. Who can access the server

   b. Who can access the database

   c. Whether the user can delete the database

   d. Whether the user can create the database

2. For added security, Shane wants to rename the Default group. This will prevent users from using the Default group to gain higher access to the database than what they should have.

   a. True

   b. False

3. Bill wants his users to be able to read and create documents. He wants users to be able to create personal agents and views. What level of access does he need to give them in the ACL?

   a. Reader

   b. Editor

   c. Designer

   d. Author

4. Sally wants her users to be able to edit any documents in the database but not be able to create new ones. She gives them Editor access to the database and removes the Create Documents option. This accomplishes her goal.

   a. True

   b. False

5. Which of these abilities is only granted to users with Manager access:

   a. Delete the Database

   b. Delete Documents

   c. Change the ACL

   d. Encrypt the Database

6. Chad sets the View access list for the Test view to allow Joe, a user with Depositor access to the database, to read the documents in that view. How does he do this?

   a. He puts Joe's name in the View Access list.

   b. He puts a group that contains Joe in the View Access list.

   c. He cannot do this. The View access list cannot override the ACL for the database.

   d. He puts the user role Depositor into the View Access list.

7. Barry has Editor access to the database, but he cannot open the Managers Only view. Why not?

   a. The view is hidden.

   b. He is excluded on the View access list.

   c. He is not in the Viewers field.

   d. He is not in the Readers field.

8. Rachel is not in the Form access list for Default Read access for documents created with this form. She has read access to the database, but she is not in the list of users that can create documents with this form. However, she can read the documents created with this form. How?

   a. She is in the View access list as a reader.

   b. She can't read the documents. It's a trick question.

   c. She is in a Names field on the form.

   d. She is in a Readers field on the form.

9. James has Editor access to the database. He is on the Read access list for a form but not on the Create access list. Can he create documents or not?

   a. Yes

   b. No

10. Janet is on the Create access list for a form. She is an author in the database. She is on the Read access list for the form and the view. However, there are documents that she cannot read. Why?

11. Connor created a database that contained some sensitive information about salaries. He created a View access list to prevent users from seeing the information visible in the Salaries view and created a Form access list that prevents users from creating documents with that form unless they are in the Human Resources department. Is this enough security?

    a. Yes

    b. No

12. Tafi creates a form with both a Readers and Authors type of field. She puts Joe in the Authors field, but she forgets to put him in the Readers field. He has Author access to the database. What access does he have to the documents created with the form?

    a. He can read the documents created with the form but not edit them.

    b. He cannot read or edit documents created with the form.

    c. He can create but not read or edit documents created with the form.

    d. He can create, read, and edit documents created with the form.

13. Bo puts an editable Authors field on a form. His users can then decide which users can edit documents they create with this form. What options can he give them for choosing the editors?

    a. Address dialog

    b. ACL dialog

    c. Any view in this database

    d. Any view in another database

14. How can you deny access to data at the field level?

    a. Use Field access lists.
    b. Use the Field properties, Deny Access to: option.
    c. Sign the field.
    d. Encrypt the field.

15. Encryption is the same for mail messages and other types of documents.

    a. True
    b. False

16. You can only encrypt fields of the following types:

    a. Text
    b. Rich Text
    c. Keywords
    d. Computed

17. You can view your secret encryption keys in the following place:

    a. File . . . Tools . . . Encryption Keys
    b. Create . . . Design . . . Encryption Keys
    c. File . . . Tools . . . UserID
    d. File . . . Database . . . Encryption Keys

18. Geoff wants to force the Salary field on the User Profile form to be encrypted. How can he do this?

    a. He can't. Encryption cannot be enforced; it is always a user option.
    b. He can choose a Default Encryption key for the form; all encrypt-enabled fields will then automatically be encrypted.
    c. He can choose a Default Encryption key for the database; all encrypt-enabled fields will automatically be encrypted.
    d. He can choose a Default Encryption key for the fields; the fields will automatically be encrypted.

19. What is the name of the reserved field required for Optional encryption on a form?

    a. EncryptionOptions
    b. EncryptOptions
    c. SecretEncryptionKeys
    d. EncryptionKeys

20. Is the encryption key called Secret the same as the encryption key called SECRET?

    a. Yes
    b. No

21. Danny creates three encryption keys. How can he distribute these keys to the users of his Instructor Evaluation database?

    a. He can attach the keys to the database by choosing Database Properties, Default Encryption keys.
    b. He can send the encryption keys using Notes mail.
    c. He can save the encryption keys into one .KEY file and distribute that.
    d. He can save the encryption keys into three .KEY files and distribute those.

22. Every user's Public key is the same. It is stored in the Public Name and Address Book. Every user's private key is different. It is stored in the user ID file.

    a. True
    b. False

23. To allow multiple users to sign a form, Polly needs to:

    a. Create multiple fields on the form that are signing-enabled.
    b. Create multiple fields that are signing-enabled and place them in a layout region.
    c. Create multiple fields that are signing-enabled and place them in separate sections.
    d. Create multiple fields that are signing-enabled and place them in one controlled access section.

24. It doesn't matter if a signing-enabled field is in a section or just on the form, signing always occurs when the document is saved or refreshed.

    a. True
    b. False

25. The option to prevent printing/forwarding/copying is very tight security. There is no way to get the data out of the document without writing it down and typing it into another application.

    a. True
    b. False

Answers

1. Both b and c are correct. The ACL determines who can access the database and what functions they can perform on the database.
2. b. False. The Default group cannot be renamed or removed.
3. d. Author, with the Create personal agents and Create personal folders/views options.
4. b. False. You cannot remove the Create documents ability.
5. a, c, and d are correct. Users with Manager access are the only ones that can delete databases, encrypt databases, set replication settings for databases, and change the ACL for databases.
6. c. He cannot do this. The View access list cannot override the ACL for the database.
7. b. He is excluded on the View access list. The View access list can limit users from seeing what it is in a view even when they have higher access.
8. d. She is in a Readers field on the form.
9. a. Yes. He can still create documents because he is an editor with at least Read access to the form.
10. The author of the document edited the Document security options and removed Janet from the list for that document.
11. b. No. A user with Editor access to the database creates a private view and can see all the information. He should have limited the Form Create access list and used Hide-When formulas as well.
12. d. He can create, read, and edit documents created with the form.
13. All the choices are available.
14. d. Encrypt the field.
15. b. False. Mail messages use Public key/Private key encryption; Field encryption on other documents uses Secret Key encryption.
16. All the answers are correct. All field types can be encrypted.
17. c. File . . . Tools . . . UserID; The Encryption keys to which you have access are stored in your user ID file.
18. b. He can choose a Default Encryption key for the form; all encrypt-enabled fields will automatically be encrypted.

19. c. SecretEncryptionKeys.
20. b. No. Encryption keys are case-sensitive.
21. b or c. He can either click the Mail button on the Encryption tab in his user ID dialog box to mail the encryption keys to the appropriate users or he can click the Export button to save each of them in a separate .KEY file and give those files to the users.
22. b. False. Although the Public key is stored in the Person document in the Public Name and Address book, every user's Public key is different and unique.
23. c. Create multiple fields that are signing-enabled and place them in separate sections.
24. b. False. When a signing-enabled field is not in a section, it is signed when the document is mailed. When a signing-enabled field is in a section, it is updated when the document is changed, saved, or refreshed.
25. b. False. Screen capture utilities can copy the data, although the data cannot be copied to the clipboard.

APPENDIX A

Level 1 @Functions

The @Functions listed in this appendix are most likely to be on the Application Development I exam. You will probably not see all of these @Functions on the exam, and you may see some not on this list, although that is unlikely. Nevertheless, these are the ones with which you should be familiar for the exam. You may be asked to pick the result of a formula that uses one or more of these functions. You might also be asked to choose the reason why one of these functions is not working.

You can find more information about each of these functions using the Notes Help database @Function reference or the official Lotus Notes documentation, *Programmer's Guide, Part 2*.

<table>
<tr><th>@Function and Syntax</th><th>Notes</th><th>Example</th></tr>
<tr><td>@Adjust (DateToAdjust; years; months; days; hours; minutes; seconds)</td><td>Adjusts date values by the noted increments; increments can be positive or negative; includes all parameters, even if they are null.</td><td>@Adjust ([11/22/95];2;-14; ;;;) returns 11/08/97.</td></tr>
<tr><td>@All</td><td>Returns a true value.</td><td>SELECT @All selects all documents for a view or agent.</td></tr>
<tr><td>@AllChildren</td><td>Returns all response type documents at all levels for any of the parent documents selected.</td><td>SELECT form = "Question" | @All Children returns all documents created with the form Question and any and all responses to those documents.</td></tr>
<tr><td>@AllDescendants</td><td>Returns all response and response-to-response type documents at all levels for any of the parent documents selected.</td><td>SELECT form = "Question" | @AllDescendants returns all documents created with the form Question and any and all responses and response-to-responses.</td></tr>
<tr><td>@Attachments</td><td>Returns the number of file attachments within the selected document.</td><td></td></tr>
<tr><td>@Command ([Command]; parameters)</td><td>Executes the specified Notes command.</td><td>@Command([FileCloseWindow]) closes the current Notes window.</td></tr>
</table>

@Function and Syntax	Notes	Example
@Contains (string; substring)	Returns True (1) if the value of the substring is contained in the string; returns False (0) if the value of the substring is not contained in the string.	@Contains ("I'm a teacher and writer"; "writer") returns the value true (1).
@Created	Returns a date and time stamp of the time when the document was created.	Returns 01/01/99 12:32:50 PM for a document created Jan. 2, 1999 at that time.
@Day (TimeDateValue)	Returns the numeric value of the day of the month in the Time/Date value.	@Day(@Created) returns 8 if the document is created on November 8.
@Do (expression1; expression2 . . .)	Returns the value of the last expression in the list; evaluates all expressions contained in the parentheses from left to right.	@Do (@Command ([FileSave]; @Command [FileCloseWindow]).
@DocChildren OR @DocChildren (string)	Returns either the numeric value of the number of response documents to the current document or the specified string, substituting the correct numeric value for the percent sign (%).	@DocChildren returns 3 if there are three response documents to the selected document; @DocChildren ("There are % Responses.") returns *There are 3 Responses.*

continues

@Function and Syntax	Notes	Example
@DocDescendants OR @DocDescendants (string)	Returns either the numeric value of the number of responses or response-to-response documents to the current document, or the specified string, substituting the correct numeric value for the percent sign (%).	@DocDescendants returns 3 if there are three response or response-to-response type documents to the selected document; @DocDescendants ("There are % Responses.") returns *There are 3 Responses.*
@DocFields	Returns the list of all fields on the current document.	
@Elements (list)	Returns a numeric value of the number of entries in the included list.	@Elements ("Libby": "Scott": "Brian": "Mike") returns 4.
@Explode (textstring; separator; includeEmpties) OR @Explode (daterange)	This function returns either the individual words or clauses of the string, if using a text string, or individual days, if using a date range. Note that the default display separator is a semicolon (;).	The function @Explode (Authors) returns Libby; Scott; Brian; Mike, if the contents of the Authors field are "Libby Scott Brian Mike".
@Failure (string)	For use in @If functions that are used in Input Validation events to determine if a particular value does not satisfy the validation criteria. Returns the value of the string you supplied.	

@Function and Syntax	Notes	Example
@Hour (TimeDate Value)	Returns the numeric value of the hour of the day in the Time/Date value.	@Hour (@Created) returns 8 if the document is created at 8:35 p.m.
@If (condition1; action1; condition2; action2; else action)	The @If statement evaluates the given conditions and then returns the specified action. If none of the statements apply, then the negative (else) action is returned. @If statements always have an odd number of arguments to allow the negative statement. @If statements can be nested. Up to 99 arguments can be used in an @If statement.	@If (Car = "Miata"; "Mazda"; Car = "Z3"; "BMW"; "This car is not defined.") If the field Car contains Tercel, this formula returns the text string "This car is not defined."
@Integer (number)	Returns a truncated whole number value for the number or number list; also removes the decimals without rounding.	@Integer (26.7778) returns 26.

continues

@Function and Syntax	Notes	Example
@IsMember (textlist; textlist) or @IsNotMember (textlist; textlist)	These functions, which are case-sensitive, return a True or False value to indicate if a piece of text or a text list is (or is not) contained within another text list.	The function returns a "1" if the word Author is in the JobDescription field: @IsMember("Author"; JobDescription).
@IsNumber (value)	Returns either (1) True or (0) False after examining a field or other value to determine if it is a number.	@IsNumber (OrderDate) returns 1 if the field OrderDate is a number field.
@IsText (value)	Returns either (1) True or (0) False after examining a field or other value to determine if it is text.	@IsText (OrderDate) returns 0 if the field OrderDate is not text.
@IsValid	Returns either (1) True or (0) False after running all Input Validation formulas on a form to determine if they all resolve to True.	
@Left (string; n), where n is the number of characters to return	Returns the appropriate number of characters, counting from the left.	@Left ("Total Seminars"; 5) returns Total.
@Length(argument)	@Length determines the number of characters in a text string.	@Length ("Total Seminars") returns 14, the number of characters in the string.

@Function and Syntax	Notes	Example
@LowerCase (string)	@LowerCase places the argument (either field, result of another formula, or text string) into all lower case. Similar @Functions include @UpperCase and @ProperCase.	@LowerCase (ColorsField) or @LowerCase ("HOUSTON"), which returns "houston."
@Max(number1; number2)	Compares two numbers or two lists of numbers and returns the highest value.	@Max (26: 130; 35: 90) returns 35 : 130.
@Member (value; list)	Returns the numeric value of the position of the value in the list.	@Member ("Libby"; "Mike": "Brian" : "Scott" : "Libby") returns 4.
@Min (number1; number2)	Compares two numbers or two lists of numbers and returns the lowest value.	@Min (26; 30) returns 26.
@Month	Returns the numeric value of the month in the Time/Date value.	@Month (@Created) returns 11 if the document is created in November.

continues

@Function and Syntax	Notes	Example
@Name ([Action]; name)	This function allows the designer to manipulate a hierarchical name of a user or server. This function is especially useful for formatting names to be displayed in views. The action parameter allows the designer to expand, abbreviate, or separate the name listed in the function. The actions allowed on a name include the following: ■ [Abbreviate]: Removes the component labels from a hierarchical name ■ [CN], [G], [I], [S], [Q], [OUn], [O], [C]: Displays the specified component of a hierarchical name. ■ [Canonicalize]: Adds the component labels to a hierarchical name, including adding any missing components. ■ [ToKeyword]: Puts the hierarchical name in a format that will allow it to be categorized in a view.	@Name ([Abbreviate]; CurrentUser) takes the value of the field CurrentUser and returns it in abbreviated format: Libby Schwarz/ Instructors/ TotalSem, rather than canonicalized.

@Function and Syntax	Notes	Example
@NewLine	Creates a carriage return in the value; cannot be used in column formulas.	"Employee\'s Last Name:" + @NewLine + LastName returns Employee's Last Name: Schwarz.
@NoteID	Returns the Unique ID number of the current document.	
@Now	Returns the current date and time.	
@PostedCommand ([Command]; arguments)	Executes the specified Notes command after all other commands and functions have been executed.	@Command([FileCloseWindow]) closes the current Notes window after all other @Commands or @Functions have been run.
@ProperCase (string)	@ProperCase places the argument (either field, result of another formula, or text string) into proper sentence case. Similar @Functions include @UpperCase and @LowerCase.	@ProperCase (UserName) or, @ProperCase ("HOUSTON"), which returns "Houston."

continues

@Function and Syntax	Notes	Example
@Replace (list; replacewhat; replacewith)	This function replaces the items in the first argument (list) that are also in the second argument (replacewhat) with the specified items in the third argument (replacewith). The items in the lists are used in order, so that the first item in the replacewhat list is replaced by the first item in the replacewith list.	@Replace ("Joe" : "Aaron" : "Kat" : "Allie"; "Joe" : "Allie" ; "John" : "Mel") replaces the names Joe and Allie from the list with the names John and Mel.
@Right (string; substring) OR @Right (string; numcharacters)	Returns characters from the right of the given string. You can specify either a number of characters or you can return all characters to the right of the substring.	@Right ("Libby is cool";4) returns cool.
@Round (number; factor)	Rounds the indicated number to either the nearest whole number or the designated factor.	@Round (26.33339) returns 26.
@Select (number; values)	Returns the value that appears in the position indicated by the number.	@Select (2; "Libby"; "Mike"; "Brian"; "Scott") returns Mike.
@SetField (FieldName; value)	Sets a field to the specified value; similar to the FIELD keyword.	@SetField (Instructor; @UserName) sets the value of the Instructor field to the current user.

@Function and Syntax	Notes	Example
@Sign (number)	Returns -1 for negative numbers, 0 for zero, and 1 for positive numbers.	@Sign (-64) returns -1.
@Sqrt (number)	Returns the square root of the given number.	@Sqrt (9) returns 3.
@Subset (list; number)	@Subset returns the appropriate number of members of the list.	@Subset ("Brian"; "Libby"; "Mike"; "Scott"; 2) returns Brian; Libby @Subset ("Brian"; "Libby"; "Mike"; "Scott"; -2) returns Mike; Scott.
@Success	For use in @If functions that are used in Input Validation events to verify that a particular value satisfies the validation criteria. Returns the value true (1).	
@Sum (numbers)	Returns the value of the sum of the numbers or number lists in the argument.	@Sum (1 : 3 : 7) returns 11.
@Text (argument)	@Text is used to convert other types of constants or variables to text. This is used since only one type of operand or variable can be used in a formula. Use the Notes Help database to examine the time-date components of @Text.	@Text (Total), where the field Total is a numeric value, converts Total to a text value.

continues

@Function and Syntax	Notes	Example
@Text (string; FormatComponent)	Often used in Input Translation formulas to convert a number or date to a text string; has optional time-date components, some of which are listed below: ■ D0: Year, month, and day. ■ D1: Month and day, year only if not the current year. ■ D2: Month and day. ■ D3: Year and month. ■ T0: Hour, minute, and second. ■ T1: Hour and minute. ■ S0: Date only. ■ S1: Time only. ■ S2: Date and time.	@Text (@Now; D2S0) returns 11/08 if the current date is November 8 in any year.

@Function and Syntax	Notes	Example
@TextToNumber (string)	Often used in Input Translation formulas to convert a text string to a number to ensure that the same data type is being used.	
@Today	Returns the current date; this is the same as @Date (@Now).	Returns 11/08/97 if today is November 8, 1997.
@Trim (string)	@Trim does not remove spaces internal to words, only those before, after, and between.	@Trim (CarName) or, @Trim ("Libby Schwarz") would return Libby Schwarz.
@UpperCase (string)	Often used in Input Translation formulas to capitalize the text string.	@UpperCase (TitleField) returns MS WORD QUESTION if that is the value of the TitleField field.
@UserAccess (server : file)	Given a file and server name, returns the current user's access level to the specified database, numbered 1-6, starting with Depositor.	
@UserName	Returns the current server or user name.	
@V2If	Similar to @If; use only for backward compatibility with Release 2.	

continues

@Function and Syntax	Notes	Example
@V3UserName	Similar to @UserName; returns the name of the current user or server in abbreviated format (if hierarchical); for use with Version 3 of Notes for backward compatibility.	
@Version	Returns the build number of the version of Notes that you are currently running.	Build 147 is version 4.6.1a.
@ViewTitle	Returns the name of the current view, including aliases (in a text list).	
@Weekday (TimeDateValue)	Returns the numeric value of the day of the week; Days are numbered 1-7, starting with Sunday.	@Weekday ([11/08/97]) returns 7.
@Year (TimeDateValue)	Returns the numeric value of the year in the Time/Date value.	@Year (@Created) returns 1971 if the document is created on November 8, 1971.
@Zone OR @Zone (TimeDateValue)	Returns the time zone setting of the current machine in terms of how many hours must be added to convert to Greenwich Mean Time; requires a time or it returns a null value.	

APPENDIX B

Level 2 @Functions

The @Functions listed in this appendix are those that are most likely to be on the Application Development II exam. You will probably not see all of these @Functions and you may see some not on this list, although that is unlikely. Nevertheless, these are the ones with which you should be familiar for the exam. In addition, you should review and be familiar with all the @Functions listed in the previous appendix. The @Functions on the Application Development II exam will build on your knowledge of the @Functions on the Application Development I exam. You may be asked to pick the result of a formula that uses one or more of these functions. You might also be asked to choose from a list of problems explaining why one of these functions is not working.

You can find more information about each of these functions using the Notes Help database @Function reference or the official Lotus Notes documentation *Programmer's Guide Part 2*.

@Function and Syntax	Notes	Example
@Accessed	The function returns the last date that the document was accessed. This includes both last read and last edited.	Returns 11/08/97 11:23:37 a.m. if the document was last accessed on this date.
@Author	The function returns a text list of all authors of the current document. To find the values, it looks in the Authors fields, if any, From fields, if any, and the $UpdatedBy reserved field.	@Author returns Libby Schwarz/Instructors/TotalSem for a document that has no Authors field or From field, but contains this name in the $UpdatedBy field.
@Contains (string; substring)	This function searches for a substring within a string and returns a True or False value. Both arguments in the function can be either text or text lists. Note that this function is case-sensitive.	The example below returns a value of 1: @Contains ("Libby Nicole Schwarz"; "N") OR: @If (@Contains (User; "Libby"); "This is the author's opinion."; "This is not the author's opinion."). This function examines the User field for the name Libby to determine which response to print.
@DbColumn (class: "NoCache"; server: database; view; ColumnNumber)	This function returns all the values in a column of a view, folder, or non-Notes database table. Use for populating keyword fields.	The function below returns the values in column 3 of the ByProducts view of the SOFTWARE.NSF database. @DbColumn ("Notes": "NoCache"; "" : "Software.NSF"; "ByProducts"; 3")

@Function and Syntax	Notes	Example
@DbExists (server : file)	Uses the server and file name or replica ID to indicate if the specified file exists. Returns True (1) or False (0).	
@DbLookup (class: "NoCache"; server :database; view; key; FieldName) OR @DbLookup (class: "NoCache"; server :database; view; key; ColumnNumber)	Returns the value of a field or column in a view for a document, based on a unique key. Can be used for Notes and non-Notes (ODBC) databases.	@DbLookup ("Notes": "NoCache"; "ServerA": "Research\\BookTest.NSF"; "ByAuthor"; "Schwarz"; "Status"). This formula returns the value of the status field for all documents created by user with the last name Schwarz in the ByAuthor view of the BOOKTEST.NSF database on the ServerA server.
@DbManager	Returns all users, servers, and groups with Manager level access to the current database.	@DbManager returns "Libby Schwarz; Instructors; Admin; ServerA" if Libby, the groups Instructors and Admin, and the server ServerA all have Manager access.
@DbName	Returns the name of the current Notes server and database. Shows nothing for the server portion if the file is local.	@DbName returns "ServerA; Research\\BookTest.NSF" if the current database is BOOKTEST.NSF, stored in the Research subdirectory on ServerA.

continues

@Function and Syntax	Notes	Example
@DbTitle	Returns the actual title of the current database.	@DbTitle returns "Libby Schwarz" if the database is my mail file.
@DeleteDocument	Deletes the current document; used only in agents.	
@DeleteField	Deletes the current or specified field; used only in agents.	
@DialogBox ("form"; [AutoVertFit] : [AutoHorzFit] : [NoCancel] : [NoNewFields] : [NoFieldUpdate] : [ReadOnly] ; "title")	Creates a dialog box that displays the current document using the specified form.	@DialogBox ("Questions"; [AutoVertFit] : [AutoHorzFit]) displays a dialog box with the form Questions. The underlying document was created with the form StoredQuestions. Both have the field UserQuestion, which is completed on StoredQuestions when the user completes the dialog box.
@DocumentUniqueID	This function returns the 32-character document UNID (unique ID). If you use this function in a formula, you can create a link to the document specified by the ID number within the current database.	

@Function and Syntax	Notes	Example
@Elements (list)	This function returns a numerical value for the number of elements in the list.	In the function below, for example, the returned value would be four: @Elements ("Mike" : "Brian" : "Libby" : "Scott")
@Environment (VariableName; value) OR @Environment (VariableName)	This function either sets or returns an environment variable stored in the user's NOTES.INI file. Use just the VariableName parameter to return the value. Works similarly to the ENVIRONMENT keyword and @SetEnvironment function.	@Environment ("UserNumber"; @Text (SocSec)) sets the variable $UserNumber in the NOTES.INI from the SocSec field in the document. @Environment ("UserNumber") returns this value in other fields, formulas, documents and databases.
@Explode (textstring; separator; includeEmpties) OR @Explode (daterange)	This function returns either the individual words or clauses of the string, if using a text string, or individual days, if using a date range.	The function @Explode (Authors) returns Libby; Scott; Brian; Mike, if the contents of the Authors field are "Libby Scott Brian Mike". Note that the default display separator is a semicolon (;).

continues

@Function and Syntax	Notes	Example
@GetDocField (documentUNID; fieldname)	This function returns the value of the field specified in the function from the document specified by the document UNID.	
@Implode (textlist; separator)	This function works in the opposite manner of @Explode and returns a concatenated text string from a list.	The function @Implode (Authors) returns Libby Scott Brian Mike if the contents of the Authors field are "Libby": "Scott": "Brian": "Mike". If there is no separator specified, a space is used.
@InheritedDocumentUniqueID	This function returns the UNID for the parent of the current document. This function is used within response documents to define and maintain the link to the parent. To make this function work properly, inheritance must be enabled, using the Inherit Values from Selected Document form property.	
@IsAvailable (fieldname)	Returns True (1) if the specified field exists on the document and False (0) if it does not.	@If (@IsAvailable (Editor); Editor; Author) returns the value of the Editor field if it exists; otherwise, it returns the value of the Author field.

@Function and Syntax	Notes	Example
@IsDocBeingEdited	Returns True (1) if the document is being edited and False (0) if it is not. Use in buttons, Hide-When, field, and form action formulas.	
@IsDocBeingLoaded	Returns True (1) if the document is being opened into memory and False (0) if it is not. Use in Hide-When, field, and form formulas.	
@IsDocBeingMailed	Returns True (1) if the document is being mailed and False (0) if it is not. Use this to calculate the number of times a document has been mailed, or to apply a different form when the document is mailed.	
@IsDocBeingRecalculated	Returns True (1) if the document is being refreshed or recalulated and False (0) if it is not. Use this in field formulas to increment the number of times a document has been refreshed.	

continues

@Function and Syntax	Notes	Example
@IsDocBeingSaved	Returns True (1) if the document is being saved and False (0) if it is not. Use in field formulas to increment the number of times a document has been saved or to change the Readers field before and after saving.	
@IsMember (textlist; textlist) or @IsNotMember (textlist; textlist)	These functions, which are case-sensitive, return a True or False value to indicate if a piece of text or a text list is (or is not) contained within another text list.	In the example below, the function returns a "1" if the word Author is in the JobDescription field: @IsMember ("Author"; JobDescription). To make sure that the words in the JobDescription field match the case shown in the text string, you might use the formula below: @IsMember ("Author"; @ProperCase (JobDescription)).
@IsNewDoc	Returns True (1) if the document is new and False (0) if it is not. It's useful to change window titles after a document has been created. Also useful in buttons, Hide-When formulas, window titles, field formulas, and form action formulas.	@If (@IsNewDoc; "New Question"; UserQuestion) returns the text "New Question" to the window title, if the document is new, and the value of the UserQuestion field, if it is opened or after it has been saved.

@Function and Syntax	Notes	Example
@Keywords (textlist; textlist; separator)	This function returns the items from the second argument that are also found in the first argument.	In the following formula, the return value would be Lotus, Microsoft, NT: @Keywords ("Lotus Notes 5.0 will be released next year" : "Windows NT 5.0 will be released next year" : "Microsoft Exchange competes with Lotus Notes"; "Lotus"; "Novell"; "Microsoft"; "IBM"; "NT")
@MailSend OR @MailSend (SendTo; CopyTo; BlindCopyTo; Subject; Remark; Body; [flags])	Either mails the current document to the user in the document's SendTo field, or creates a new mail memo based on the values of the parameters.	@MailSend ("Libby Schwarz/Instructors/TotalSem"; "";""; "New Book Contract."; "This is the latest book contract we have received."; "ContractNumber":"BookTitle": "BookTopic") sends a mail message to Libby Schwarz with the subject New Book Contract that includes the values of the ContractNumber, BookTitle, and BookTopic fields.

continues

@Function and Syntax	Notes	Example
@Member (value; list)	This function returns a numerical value relating to the position of the value in the list.	The function below returns a value of 3 if the name Libby is the third item in the field SeniorityList: @Member ("Libby"; SeniorityList).
@Name ([Action]; name)	Allows the designer to manipulate the hierarchical name of a user or server. This function is especially useful for formatting names to be displayed in views. The action parameter allows the designer to expand, abbreviate, or separate the name listed in the function. See Appendix I for a list of actions.	@Name ([Abbreviate]; CurrentUser) takes the value of the field CurrentUser and returns it in abbreviated format, Libby Schwarz/ Instructors/ TotalSem, rather than canonicalized.
@PickList ([NAME]: [SINGLE]) OR @PickList ([CUSTOM]; server : database; view; title; prompt; column)	Displays a dialog box window that displays either the Address dialog box or the specified view. Do not use in views; use to display views.	@PickList ([NAME]: [SINGLE]) allows the user to pick a single name or group from the Name and Address Book.
@Prompt ([style] : [NoSort]; title; prompt; DefaultChoice; ChoiceList)	Displays a dialog box with the indicated parameters to the user; returns a value based on the user's input.	@Prompt ([OK]; "Please remember to replicate the database after making changes") displays that message to the user in a dialog box with an OK button provided.

@Function and Syntax	Notes	Example
@Replace (list; replacewhat; replacewith)	This function replaces the items in the first argument (list) that are also in the second argument (replacewhat) with the specified items in the third argument (replacewith). The items in the lists are used in order, so that the first item in the replacewhat list is replaced by the first item in the replacewith list.	@Replace ("Joe" : "Aaron" : "Kat" : "Allie"; "Joe" : "Allie" ; "John" : "Mel") replaces the names Joe and Allie from the list with the names John and Mel.
@Return	Stops the execution or evaluation of a formula when it reaches this function. Generally used with an @If function.	@If (@Prompt ([YesNo]; "Do you want to create another Question?"); @Command(C [Compose];"Question"); @Return ("")) The @Prompt gives the user the choice to create another question. If they choose Yes, it opens another Question form to complete it; if not, the formula stops.
@SetDocField (documentUNID; fieldname; new value)	This function sets the value of the field specified in the function with the value specified in the function of the document specified by the document UNID.	

continues

@Function and Syntax	Notes	Example
@SetEnvironment (VariableName; value)	This function sets an environment variable stored in the user's NOTES.INI file. It works similarly to the ENVIRONMENT keyword and @Environment function.	@SetEnvironment ("UserNumber"; @Text (SocSec)) sets the variable $UserNumber in the NOTES.INI from the SocSec field in the document.
@Subset (list; number)	This function moves from left to right through the list and returns the number of items specified. If the number is a negative number, the function searches from right to left.	The function below returns Brian; Scott: @Subset ("Bob" : "Ron" : "Joe" : "Brian" : "Scott"; -2).
@Unique (textlist)	The argument in this function is optional. If used without the argument, @Unique returns a random, unique text value. When used with the list function, the function returns the first occurrence of each item in the list.	

Index

Boldface folios denote illustrations.

About the Author

Libby Ingrassia Schwarz is a Certified Lotus Professional for both Application Development and System Administration. She is also a Microsoft Certified Systems Engineer specializing in Windows NT, TCP/IP, and Exchange. Her training credentials include being a Certified Lotus Instructor, Microsoft Certified Trainer, and Certified Technical Trainer.

Libby has been in the computer industry doing technical writing, training, course development, and consulting in Notes, NT, and other technologies since 1994. She has planned and implemented installations, taught classes, and developed applications for a variety of technologies. She currently writes and teaches for Total Seminars, LLC. You can send her e-mail at `LibbyS@TotalSem.com`.

Total Seminars is a training company specializing in government and corporate on-site training and training material development. Total Seminars instructors have written books and manuals on A+ Certification, Networking, MCSE Certification, and Lotus Notes. For information on books and on-site training courses, see their Web page at `www.totalsem.com`.

Application Development Exam Quick Reference Card

Remember these points during the exams, and make them the last items you run through before taking the exam. They are not the only information necessary to pass, but they give you basic ideas to memorize.

Lotus Notes Application Development I: Exam 190–271

NOTES APPLICATIONS

Notes does not support real-time data access, complex mathematical computations, complex reports and queries, or data larger than 4 GB.
Enter data on a form, which is then saved in a document. Documents are displayed and accessed through views.
Set the database size from 1 to 4 GB at creation. The default is 1 GB.
The Database Properties InfoBox Launch Tab enables you to launch a Navigator, link, or About Database document automatically.
Full-text indexes cannot search bitmap graphics or About or Using documents, but they can search text attachments and encrypted fields.

CREATING THE FORM

Use synonyms with form names by typing a pipe symbol (vertical bar) after the form name. Use synonyms internally when referring to forms, as in formulas.
Cascade form names off the create menu by using a slash between cascade choices and form names.
The three form types are: main document, response, and response-to-response. Response forms can be a child to a main document. Response-to-response forms can be a child to a response form or main document.
New document versions can be responses, parents, and siblings, as selected on the Form Properties InfoBox.
So users can be anonymous when creating forms, check the anonymous form option on the Form Properties InfoBox.
To merge changes on different document replicas, use the Merge Replication Conflicts option on the Form Properties InfoBox.
Sections can be standard or controlled-access and can contain most form elements.
Use Hide-When properties to hide sections or paragraphs on a form based on document events (such as when opened for reading or editing) or based on a formula.
Button labels determine their length.
Use subforms to reuse design elements (text, fields, sections) on multiple forms.

Fields, Formulas, Actions, and Agents

Fields can be editable or computed. If editable, they can have default value formulas, input translation formulas, and input validation formulas. If computed, they can be computed, computed for display, or computed when composed. Computed for display is calculated each time the document opens and is not stored. Computed when composed is calculated once. Computed is calculated as save, close, open, and refresh.
Field types: Text, time, number, rich text, authors, readers, names, keywords.
Checkboxes assume multi-values, and radio buttons do not enable multi-values. Dialog lists offer a choice.
Only dialog lists enable you to add values not in the list.
Shared fields enable you to share a single field among multiple document forms.
To base an input validation formula, hide-when formula, or other computed field on the keyword field value, enable refresh fields on keyword change.
Order of operation: parenthesis, exponents, multiplication, division, addition, subtraction.
Know the @If function: @If (condition; if yes; else); always an odd number of arguments, up to 99 arguments. Review the @Function table in Chapter 3.
@Commands are run in the order they exist in the form. Use @PostedCommands to run a command at the end of the form (and to be compatible with v3 forms).

VIEWS AND FOLDERS

Designers select documents for a view. End users place documents in a folder. Select documents for a view by using a selection formula or search builder.
Views and Folders can be private, shared, or shared personal on first use. Choose this option when the view or folder is created (it cannot be changed).
You can use synonyms (or aliases) for view names.
Format your views using beveled or simple headings, multiple lines per heading or row, shrinking rows to content when using multiple lines, and enabling larger row spacing for more space between rows (documents).
@AllChildren selects only direct children (responses) to a selected document. @AllDescendents selects all children and grandchildren (response and response-to-response) of a selected document. @IsResponseDoc selects all documents created with a response or response-to-response type form.
If you use synonyms for keywords when creating fields, they will be displayed in views.

COMPLETING THE DATABASE

The ACL has seven access levels: Manager, Designer, Editor, Author, Reader, Depositor, No Access.
When a database is created, the default access is Designer. The database creator has Manager access.
Create database help by editing the About this Database and Using this Database documents.
Create graphical Navigators so users can easily move through the database.
Navigators and About documents can be launched automatically using Database Properties.

Lotus Notes Application Development II: Exam 190–272

ADVANCED APPLICATIONS AND WORKFLOW

Mail-enable Notes forms by adding a MailOptions reserved field, @MailSend function, or choosing On Close: Present Mail Send dialog from the Form Properties InfoBox.
Mail-enabled Notes forms must have a SendTo field or other recipient list.
@MailSend (SendTo; CopyTo; BlindCopyTo; subject; remark; BodyFields; [flag]:[flag])
Mail-in database documents must be created in the NAB so mail can be delivered to any databases (other than users' mail files).

ADVANCED FORM ELEMENTS

Computed subform formulas are calculated at compose—they are not recalculated on field changes.
The Computed subform formula must evaluate to the name of a database subform.
Controlled-access sections do not override the ACL. They secure the data from editing, not reading.
Benefits of layout regions include layering, controlling the fields' tab order, creating dialog boxes, and controlling text placement and hide-when.
Place controlled-access sections and layout regions on a subform, but do not place a section or subform in a layout region.
Use hide-when to hide sections and layout regions.
In layout regions, keyword fields accept these interface choices: list box, combo box, check box, radio button. You cannot use dialog lists or values not in the list.

ADVANCED FIELDS, FUNCTIONS, FORMULAS, ACTIONS, AND AGENTS

Editable, Default Value, and Computed When Composed fields are evaluated at Compose Only; Editable, Input Translation fields, and Editable, Input Validation fields are evaluated at Refresh, Save; Computed fields are evaluated at Compose, Refresh, Save; Computed for Display fields are evaluated at Compose, Refresh, Save, Load but are not saved.
Internal fields are named with a $ and are stored in document properties. Know the $UpdatedBy and Form fields.
@DbColumn [@DbColumn (class: "NoCache"; server:database; view; ColumnNumber)] returns values of a view, folder, or table column—in a non-Notes database.
@DbLookup [@DbLookup (class : "NoCache"; server:database; view; key; FieldName)] returns the value of a certain field or column in a view, based on a unique, specified key.
@DbColumn functions only return 64 KB of data.
@PickList displays a dialog box so users can choose a document: @PickList ([NAME]: [SINGLE]) OR @PickList ([CUSTOM]; server : database; view; title; prompt; column).
@Prompt displays a dialog box so users can enter values: @Prompt ([style] : [NoSort]; title; prompt; DefaultChoice; ChoiceList). Refer to Chapter 8 for style information.
@DialogBox displays the current document using a different form, subform, or layout region: @DialogBox ("form"; [AutoVertFit] : [AutoHorzFit] : [NoCancel] : [NoNewFields] : [NoFieldUpdate] : [ReadOnly] ; "title").
Review multi-line and complex formulas, especially complex @IF functions.
Know simple actions available in an agent: Copy to database or folder, Delete from database, Mark documents read or unread, Modify field or fields, Move to or remove from folder, Reply to sender, and Send document, message, or Newsletter summary.

ADVANCED VIEWS

Notes uses these items to decide which form displays a document: the form stored in the document; form formula; the form that created the document; the default form for the database.
Use search builder or selection formulas to decide which documents are displayed in a view.
You can sort columns and allow end users to sort columns.
Use hidden columns for additional sorting.

MAINTAIN DATA CONSISTENCY

The *Database is a Template* property is the defining characteristic of a master or design template, as opposed to a Notes template file.
To link the database and master template, enable the Inherit future design changes checkbox on the New Database dialog box or enable the Inherit Design from Template checkbox on the Database properties InfoBox.
Distribute design changes to databases automatically by using Server Design task, manually using File . . . Database . . . Refresh Design, or manually using File . . . Database . . . Replace Design.
These items are not updated during the replace process: database icon, database title and category, database security, including database ACL and any encryption settings, About this Database and the Using this Database documents, and any elements marked Do not enable replace/refresh to update.
Use the design document to inherit the design element from a specific template or to disable replace/refresh to update.

SECURITY

View and form access lists and readers and authors fields can limit an ACL setting but not override it.
Users listed in an Authors field can read documents, whether or not they are listed in a Readers field.
Users with Editor access and higher are not affected by an Authors field.
Encryption protects field-level data.
Attachments in encrypted Rich Text fields will be visible to all document readers.
Appropriate encryption keys must be stored in a user's ID file so they can read encrypted data.
If designers apply multiple encryption keys, users need all of them to read data.
When an encryption key is applied to a document, all encrypt-enabled fields on that document are encrypted with that key.
Any field can be encryption- or signing-enabled.
If you sign a field in a section, the section is signed each time the document is saved. If you sign a field not in a section, the document is signed when mailed.
Do not sign reserved fields such as SendTo and MailTo. They may be resigned during mailing.
Use ACL User Roles in View and Form access lists and Reader and Author fields to centralize security and ensure that administrators change all access when they change the ACL.